URBAN GROWTH'S IMPACT ON GROUNDWATER SYSTEMS

R.LILLY

ABSTRACT

In most cities, water quantity and water quality depreciation are found to be the major issues that affect the environment. Due to urbanization and growth of population, water becomes the vital property which has to be guarded safe for our future growth. Hence, any development has to be done with a cautious eye on this precious source. This research focuses on the impact of urban development strategies, particularly the construction of underground Metro rail corridors in the heart of Chennai City, on the availability of water, i.e., water quantity as well as the water quality.

The construction of Metro rail corridor involving lots of pumping of water and the tunneling work affects the ground water aquifers. The change of the heterogeneous soil strata affects the water quality too. An attempt has been made to study the water availability both quantity and quality in the study area and to arrive at a holistic solution. The study aims to evaluate the environmental pollution associated with water quality parameters into the groundwater, using contaminant transport model. The work extends to study the changes in the ground water levels due to the obstruction created in the name of tunnel below the surface level as a barrier in the soil strata.

To understand the variations in the groundwater levels and water quality, spatial and temporal analysis has been done using the premonsoon and post monsoon water levels and with the water quality parameters. The comparison has been done in three phases before, during and after construction. 20 wells are identified opposite to the construction of Underground metro rail corridors on either side 5 wells each. Levels are monitored and the samples are taken for quality analysis to study the effect of tunneling on either side of the corridor. Using water table fluctuation method, the storage of the aquifer was identified for before and after the construction. In MODFLOW, this is identified and the storage is identified with the mass balance for different time steps. Groundwater flow variations and the equipotential head variations clearly show the changes in the flow direction during the Model prediction. Some of the parameters of water quality shows the changes, but have not predominantly proved that the water quality has got deterioted.

MODFLOW and MT3DMS programs in Visual MODFLOW have been used to study the groundwater flow and contaminant transport in the groundwater. In flow modeling, three scenarios have been considered for predicting the changes in the velocity direction and the ground water level. Scenario 1 and Scenario 2 do not show any predominant changes but in the Scenario 3, flow direction has got changed. Hence, it is evident that the tunnel construction behaves as a barrier and the changes in the direction of the flow make the head equipotentials higher at one end and lower at the opposite side of the tunnel. Relative to the ground water contamination, the quality parameters are getting gradually deteriorated under the three Scenarios but in a very nominal level.

The technical analysis of the groundwater flow and the socio-economic survey regarding the groundwater level and quality perception has been overplayed to determine the ground truth of the study. It is inferred that the respondents' perception on water quality changes is in agreement with the flow direction obtained from the groundwater flow model.

Finding of wells in the city for primary data and collection of bore well data is the major difficulty. Fixing the optimization parameters to represent the ground truth of the study in the model is the major challenge. Original achievement relies in finding the variation in the prediction of the ground water flow and the management strategies required for the sustainability of the ground water status in the study area

Key words: Groundwater level, water quality, socio-economic issues.

TABLE OF CONTENTS

LIST OF TABLES

LIST OF FIGURES

LIST OF SYMBOLS AND ABBREVIATIONS

APHA	-	American Public Health Association
ASI	-	Archaeological Survey of India
BWDB	-	Bangladesh Water Development Board
BOD	-	Biochemical Oxygen Demand
BRO	-	Border Roads Organization
BC	-	Boundary Condition
BIS	-	Bureau of Indian Standards
COD	-	Chemical Oxygen Demand
Cl	-	Chloride
AutoCAD	-	Commercial Computer Aided Design
DO	-	Dissolved Oxygen
EC	-	Electrical Conductivity
ET	-	Evapo Transpiration
FEFLOW	-	Finite Element Subsurface Flow and Transport Simulation System
Fl	-	Fluoride
GIScience	-	Geographic Information Science
GIS	-	Geographic Information System
GPS	-	Global Positioning System
GWPI	-	Ground Water Potential Index
K	-	Hydraulic conductivity
ICMR	-	Indian Council of Medical Research
IRS	-	Institute of Remote Sensing
IDW	-	Inverse Distance Weighted
KRCL	-	Konkan Railway Corporation Limited
LANDSAT5	-	Land Remote-Sensing Satellite
LISS	-	Linear Imaging Self Scanning Sensor

MCL	-	Maximum Contaminant Level
MSL	-	Mean Sea Level
MLD	-	Million Litres per Day
MT3D	-	Modular 3-D Multi-Species Transport Model
MODFLOW	-	Modular Finite-Difference Ground-Water Flow Model
MODPATH	-	Modular Path
NHAI	-	National Highway Authority of India
NE	-	North East
pH	-	Potential Hydrogen
PWD	-	Public Works Department
S_y	-	Specific yield
SI	-	Sub Index
TM	-	Thematic Mapper
TDS	-	Total Dissolved Solids
TH	-	Total Hardness
TBM	-	Tunnel Boring Machine
USGS	-	United States Geological Survey
USPH	-	United States Public Health
ΔH	-	Water level fluctuation
WQI	-	Water Quality Index
WHO	-	World Health Organisation

CHAPTER 1

INTRODUCTION

1.1   GENERAL

Groundwater is an essential natural resource used by agricultural, industrial and residential sectors for irrigation, manufacturing and drinking water purposes. Water is not an inexhaustible resource and proposed actions need to be evaluated for their potential impact on both the quantity and the quality of the groundwater. Once a groundwater supply is exhausted or contaminated, it is very expensive and sometimes impossible to replace. To maintain a sustainable system, the amount of water withdrawn from these groundwater sources must be balanced with the amount of water returned to the groundwater source. This is known as groundwater recharge.

Hydrogeology determines the areas that have sufficient groundwater or recharge rates based on the soils, topography and distance between well site and land fill. Groundwater withdrawal results in significant reduction in storage or flow in a nearby surface water body. The quality and the quantity of the local aquifer systems get always altered due to urbanization process in various ways. Because of urbanization, the hydrological cycle gets changed and it affects the local water resources in terms of flooding and water shortage, in turn the changes in the ground water regimes occur.

The point and non-point sources of contaminants are generated abundantly in urban areas. The point sources are likely to affect the underground storage facilities and groundwater quality. The natural recharge of aquifers gets affected by the rapid growth of urban area due to sealing effect of the ground with concrete. The pollution of groundwater occurs due to leakage from drainage and industrial wastage and effluents. Hence, the protection of water quantity and quality becomes a major challenge of water resources research, particularly in ground water.

1.2 GROUND WATER OCCURRENCE

Ground water below the ground surface is divided into four zones soil zone: intermediate zone, capillary zone and saturated zone. Groundwater occurs everywhere beneath the Earth's surface, but is usually restricted to depth less than about 750 meters. The molecular attraction between the water, rock and earth materials and the attraction between the water particles are the two forces which prevent the ground water from moving downward. The zone of aeration is further divided into vadose zone consisting of three layers: soil moisture zone, intermediate zone, and capillary zone. It is followed by phreatic zone. The water available in this zone is called ground water.

The zone of aeration and the zone of saturation are divided by the ground water table. Figure 1.1 shows the various ground water zones. The water gets infiltrated from the surface and enters the zone of aeration through the zone with reduced permeability. The water also enters the aeration zone through the process of capillary fringe from the saturation zone.

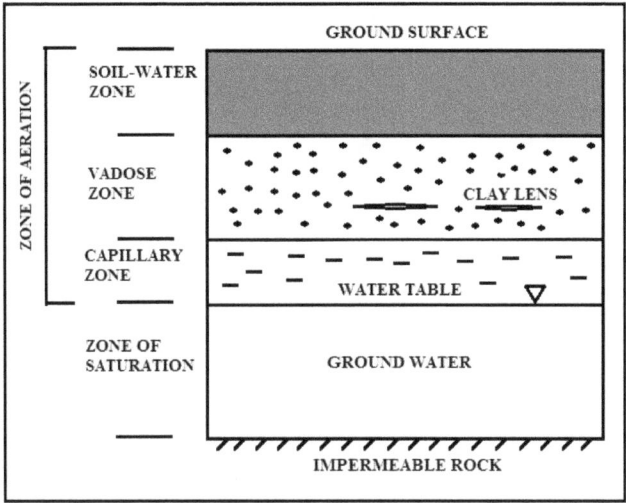

Figure 1.1 Ground Water Zones

1.3 GEOLOGICAL FORMATION OF GROUND WATER REGIME

A geological structure that is fully saturated with water and yielding sufficient amount of water is known as aquifer. Aquifer is further classified into confined aquifer and unconfined aquifer. Geological formations are further categorized into aquicludes, aquitards and aquifuges. Aquicludes are small saturated layers above the impermeable layers such as clay and shale.

Aquitards are confining layers but cannot completely check water flow to or from an adjacent aquifer. An aquifuge consists of a rock layer which has no interconnected opening or interstices. So, it neither stores nor transmits water such as Quartzite and Obsidian. Figure 1.2 shows the geological formation of various kinds of aquifers.

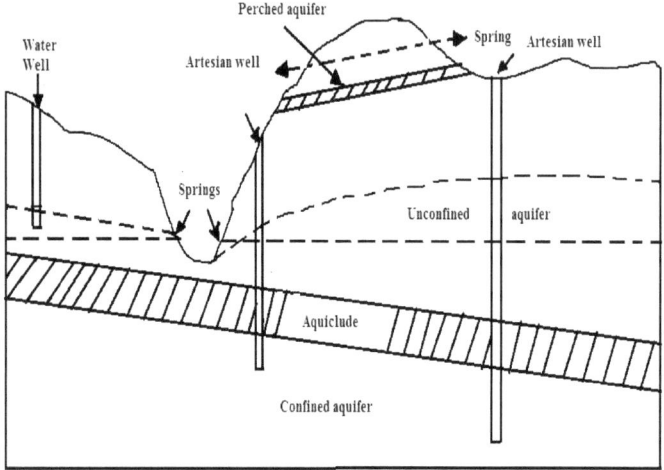

Figure 1.2 Geological formations of Aquifers

The occurrence of groundwater is influenced by the following factors. In arid regions, groundwater is easily available at great depths and in humid regions it exists at shallow depths. Ground water table increases during the rainy season and decreases in dry season. Ground water availability depends upon the topography and it plays a major role for its existence. Also, it is found to be higher near the hill tops and lower near the valleys. The descending of water table occurs due to the seepage of water into streams and lakes.

The storage and movement of groundwater is greatly affected by porosity and permeability of the underground materials. The pore space and the fractures present between the mineral grains determine the volume of water that can be retained as ground water. Permeability decides the transmission of water below the surface of ground.

1.4   GROUND WATER DYNAMICS

A groundwater system comprises the subsurface water, the geologic media containing the water, flow boundaries and sources (such as recharge) and sinks (such as springs, interaquifer flow or wells). Water flows through and is stored within the system. Under natural conditions, the travel time of water from areas of recharge to areas of discharge can range from less than a day to more than a million years. Water stored within the system can range in age from recent precipitation to water trapped in the sediments as they were deposited in geologic time.

Dynamic (renewable) groundwater resources depend on annual recharge of rainfall in the system. Groundwater flow occurs due to the elevation differences, that is from the high pressure area to the low pressure area. The difference in energy between two points that are l metre apart horizontally on a sloping water table is determined by the difference in height (h) between them. This height is called the head of water. The slope of the water table is called the hydraulic gradient and is defined as h/l. The rate of groundwater movement is related to the hydraulic gradient by Darcy's law

$$Q = KAh/l \qquad\qquad (1.1)$$

where

Q   -   Volume of water flowing in unit time (m^3/s)

K   -   Permeability(m/s)

A   -   Area of cross section of flow(m^2)

h   -   Difference in height(m)

l   -   Length of flow(m)

The hydraulic conductivity depends on the properties of the rock that allows water to flow through it (its permeability) and also on the properties of the water. Unlike hydraulic conductivity, permeability is an intrinsic property of the rock, so it is the same whatever the nature of the fluid flowing through the rock - whether water, as in this instance, or oil or gas. The hydraulic conductivity (K), however, depends on the density and viscosity of the fluid, so it will vary accordingly.

When the fluid is water, the most important factor that affects the hydraulic conductivity is temperature. For example, an increase in water temperature from 5 °C to about 30 °C will double the hydraulic conductivity and, from Darcy's law, will therefore double the speed at which the groundwater flows. Figure 1.3 explains the flow of water through a permeable rock as per Darcy'slaw below the water table for the length of flow, l and with head of water, h.

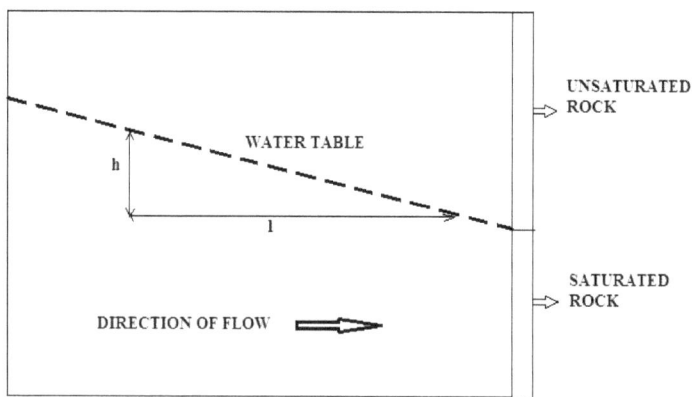

Figure 1.3   Flow of water through a permeable rock below the water table

The movement of ground water through pore spaces occurs at a very slow velocity. Hydraulic head occurs due to the differences of water table. The water percolates from the high water table area to low area of water table due to gravity. Since the permeability decreases downwards due to the compacted overlying soil, the infiltration in the vertical direction decreases.

In shallow depths, the nature of groundwater acts both as reservoir as well as conduit. The saturated zone gets recharged as the precipitation adds up the water, then the water is moved to the discharge areas like river and other water bodies. In India which experiences the monsoon climate, the water table levels and increases and show high water table beneath hills and during dry periods it decreases to the level of valleys.

1.5   SEA WATER INTRUSION

The natural processes or a human activity which allow the movement of seawater into fresh water aquifers is called Seawater intrusion. The decrease in groundwater levels or rise in seawater levels becomes the cause for the seawater intrusion. Pumping out of fresh water rapidly from the source lowers the height of fresh water in the aquifer, thus forming the cone of depression. Groundwater extractions are the primary cause of saltwater intrusion. Groundwater extraction can lower the level of the freshwater table, reducing the pressure exerted by the freshwater column and allowing the denser saltwater to move inland laterally reducing the water table to below sea level and causing widespread intrusion and contamination of water supply wells. Figure 1.4 explains the process of salt water intrusion.

The over extraction of ground water table creates a cone of depression at the sea level and the same pressure creates a cone of ascension from the fresh water which facilitates the salt water intrusion to the fresh water table.

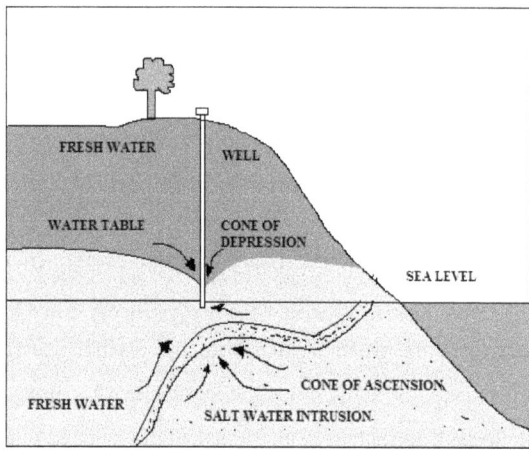

Figure 1.4 Process of Sea Water Intrusion

Groundwater extraction can also lead to well contamination by causing upwelling, or upcoming, of saltwater from the depths of the aquifer. Under baseline conditions, a saltwater wedge extends inland, underneath the freshwater because of its higher density. Water supply wells located over or near the saltwater wedge can draw the saltwater upward, creating a saltwater cone that might reach and contaminate the well. Some aquifers are predisposed towards this type of intrusion, though a relatively impermeable rock or clay layer separates fresh groundwater from saltwater, isolated cracks breach the confining layer, promoting upward movement of saltwater. Pumping of groundwater strengthens this effect by lowering the water table, reducing the downward push of fresh water. The construction of canals and drainage networks can lead to saltwater intrusion. Canals provide conduits for saltwater to be carried inland, as does the deepening of existing channels for navigation purposes. Drainage networks constructed to drain flat coastal areas can lead to intrusion by lowering the freshwater table, reducing the water.

The intrusion affects the pumping well sites in terms of quality in the undeveloped portions of the aquifer. The salt gets washed off from the surface and percolates deep into the underground aquifers. High concentrations of chloride make the water unfit for drinking and create health related problems. The concentration of sodium ions increases the problem of heart diseases and high blood pressure to the susceptible individuals.

1.6   SUB -SURFACE DEVELOPMENTS

Underground development is an important phenomenon in developing and reshaping urban areas to meet the challenges of the future. Placement of infrastructure and other facilities underground presents an opportunity for realizing new functions in urban areas without destroying heritages or negatively impacting the surface environment (Wout Broere 2016).

Urbanization occupies a puzzling position in country development and growth, global economic development, energy consumption, natural resource use, and human wellbeing (McDonald *et al.* 2011, 2014; Uddameri *et al.* 2014; Mondal *et al.* 2015; Pandey & Joshi 2015; Jain *et al.* 2016). Some of the underground developments around the world are given below.

The underground railway station in George street, Sydney shows a massive construction built underground for a distance of 1.18 km with two platform levels with a depth of 6m (20ft) and 12m (39 ft) in the upper level. Carleton University, Ottawa, Ontario, Canada is an extensive system of underground tunnels which extend upto a depth of nearly (40 ft) links the buildings of the campus, such that the members of the university need not walk outside while moving across campus.

South India's first underground metro station in Bangalore constructed for 4.8 km long underground section from Cubbon Park to City railway station completes the 18.10 km long east west corridor, also called Purple Line.

Singapore builds the largest underground substation which extends to three hectares below the ground. The facilities include Electrical sub stations for providing electricity to estates which they are connected with underground cable network. Network of tunnels that operates on gravity transports sewage and waste water to two centralized water reclamation points. The other facilities like pedestrian links, waste disposal, common service tunnel and most efficiently designed underground water reservoirs are erected below in the underground substation.

Infrastructure developments such as tunneled metro rail corridor often interfere directly with water resources. The changes due to the subsurface structure cause numerous anthropogenic impacts making urban geological and hydro geological issues complex (Huggenberger & Epting 2011). During tunneling, the trenches are excavated and to keep it dry groundwater should be pumped out from its bottom part. This causes lowering of groundwater table in the surroundings of the trench (Aivars Spalvins *et al.* 2012). Due to tunneling, the potential to cause groundwater drawdown and the ground settlement that may affect existing structures will be an issue (Sian France *et al.* 2010).

1.7   IMPACT ON WATER QUANTITY   .

Urbanization effect on water quantity results in changes in flow pattern, ground water drawdown and the deepening of water table. In turn, the reduction of natural discharges and the outflow decreases and to some extent land subsidences occur.

Urban development causes a reduction in aquifer recharge due to the impermeable seal created across   large swaths of land surface by asphalt, concrete and roof- topped structures.

Deep well drilling is the another method that causes major changes in the ground water regime. The high rate of ground water withdrawal and the decreasing recharge results in poor sanitation. The obstruction that causes changes in the direction of the flow and the water table reduction are shown in Figure 1.5.

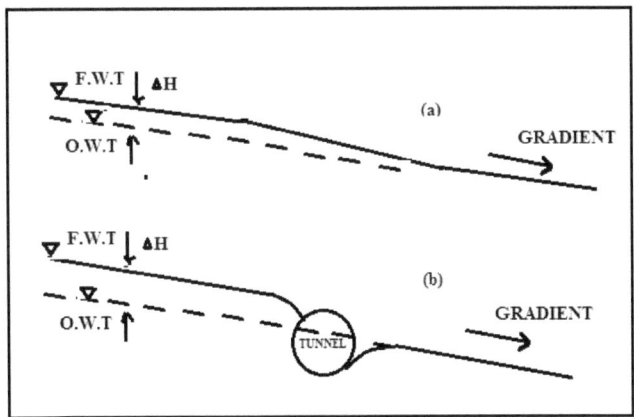

Figure 1.5   Change of Ground Water Regime due to Tunneling

1.8   IMPACT ON WATER QUALITY

The barrier effect of the tunneling below the ground brings down the changes in the flow pattern affecting the ground water quality distribution below the ground surface. Displacement of salt waters, including sea water intrusion occurs due to the underground development. Urban pollutant sources are normally classified into "point" sources that originate at a very specific location, and "line" or "distributed" sources, which tend to have a much more widespread impact on the aquifer.

Figure 1.6 explains the process of contamination of ground water quality. The oxygen from the atmosphere, air borne salts from precipitation, dissolved salts from runoff and the salinity contaminants due to human effects are the main reasons for ground water quality contamination. The soil contaminants affect the process of recharge and enter the ground water affecting its quality.

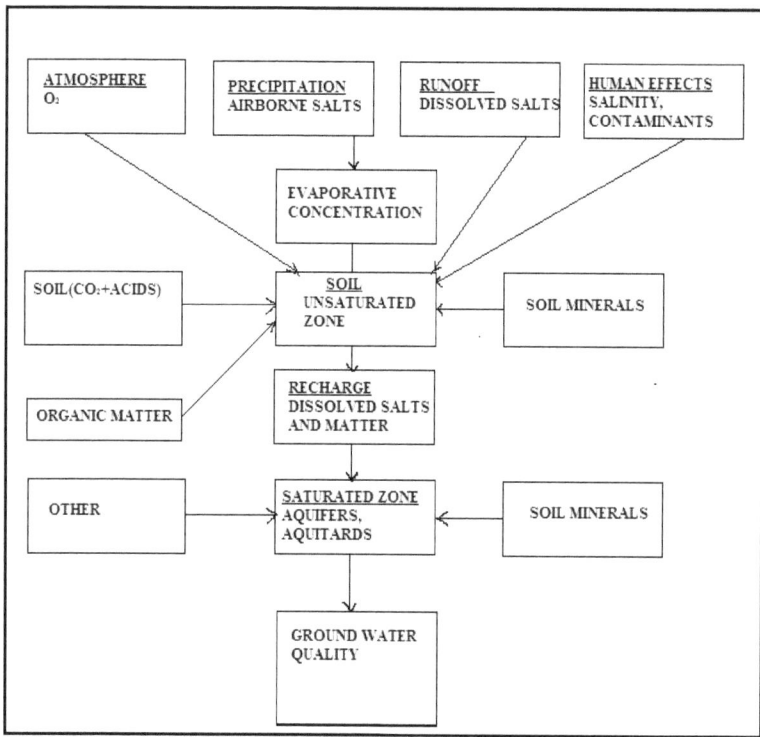

Figure 1.6 Contamination of Ground Water Quality

1.9   NEED FOR THE RESEARCH

Rapid growth of the population and the large scale urbanization to meet the demands of the people are the simultaneous, inevitable happenings that form the present scenario of the nation. This unprecedented growth of population expects a greater demand of good infrastructure facilities, availability of water for both domestic and industrial expansions, good quality of water for drinking, electricity and good drainage facilities.

The water need is the predominant one and becomes the basis for all the developments. Hence, conservation of water and preserving the ground water bodies becomes the essential need for the sustainable development of the nation.

Urbanization is the predominant global phenomenon of our time. Ground water is an important component of water resource systems, but is a critical, unappreciated, resource for urban water supply. It is extracted from aquifers through pumping wells and supplied for domestic use, industry and agriculture. With increased withdrawal of ground water, the quality of groundwater has been continuously deteriorating.

Figure 1.7 illustrates the problem of ground water lowering and contaminant issues arising due to the population growth and the urbanization need for the growing population. It also shows that the construction operation of the tunneling process becomes the major cause for the ground water lowering and contamination.

Hence, an interdisciplinary research will bring out a meaningful, acceptable and sustainable solution for these problems. This study makes an attempt to collaborate different disciplines like water resources, environment, geology, soil, chemistry with social components. This interdisciplinary study will bring out the exact truth of the ground water issue and the social concepts.

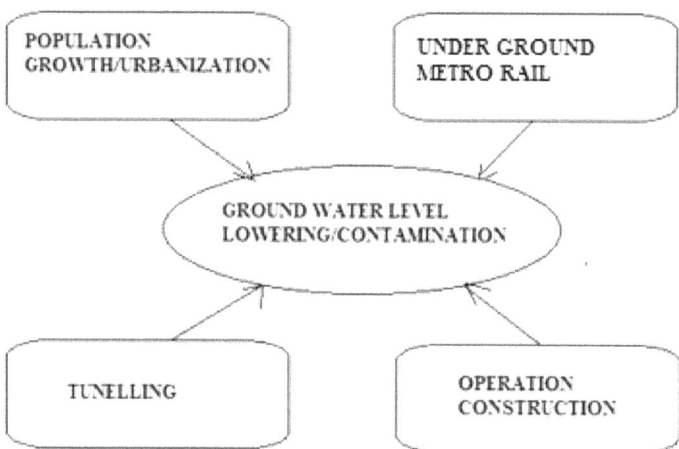

Figure 1.7 Groundwater Level Lowering/Contamination issues

1.10   OBJECTIVES

This study primarily aims to analyze the ground water regime in terms of quantity and quality before and after a subsurface development. The specific objectives of the present study are:

- To assess the water quantity and quality variations through spatial and temporal analysis.

- To develop a conceptual model representing the hydro geological condition of the well field based on ground observations and data analysis.

- To predict the ground water flow and transport due to the urbanization stress.

1.11   ORGANIZATION OF THE THESIS

The thesis consists of Seven chapters. This **Chapter 1** presents the introduction towards ground water occurrence, geological formation, ground water dynamics and the impact on both water quantity and quality. **Chapter 2** presents the literature review with respect to groundwater declination, GIS, water quality and MODFLOW. The review highlights the earlier works done in groundwater quality, spatial and temporal analysis of ground water level, socio – economic issues and modeling of groundwater flow.

Chapter 3 describes the details of the study area and the raw data collected related to describe the features of study area, water level and water quality datas.

The methodology adopted to analyze the issue of groundwater level declination and the declined water quality is presented in **Chapter 4.**

Chapter 5 explains the impact of ground water quantity **Chapter 6** discusses the impact of ground water quality.

Summary of the research findings and recommendations to reduce the water level declination and contamination and management practices to address socio-economic issues are given in **Chapter 7.**

CHAPTER 2

REVIEW OF LITERATURE

2.1   GENERAL

Groundwater level changes and the declination of ground water quality due to tunneling becomes a major issue, since it affects the environment and the development of the society and social well-being. The review starts with the study of underground developments in the National and International levels. A deep understanding is needed to carry out the analysis for a truthful solution regarding the ground water declination and the water pollution. This includes methods of tunneling and their effects on water table, characterization of water quality parameters with respect to tunneling operation, lithology variation, groundwater flow and contaminant transport modeling and social issues of the people. Literature relative to the above have been reviewed and given below.

2.2   STRUCTURAL UNDERGROUND DEVELOPMENTS

Increase in population and the technology developments lead to underground construction for many purposes such as shops, parking lots, mass rapid transport etc.(transmit system)underground developments in rapid rate. Underground developments in the form of railway stations, temples, houses, storage yards, tunnels and hotels are a few concrete examples of underground sub structures. The fabulous underground developments are segregated into National and International level and they are given in the following sections.

2.2.1 National Development

Palika Bazar, New Delhi, India is an underground market constructed as a part of the Connaught place shopping. It is a cut-and-cover subsurface structure with a beautiful garden created above it (Goel 2015).

In ancient India, underground structures existed as dwelling pits cut into the compacted loess deposits in Kashmir around 3000 B.C. and 500 B.C. During excavations in 1960, it was brought to light by the Archaeological Survey of India (ASI). These pit houses were found to provide excellent protection against cold and severe winter as well as heat of summer. It also offered protection against external attack. Dwellings belonging to 1600 B.C. were also discovered at Nagarjuna Konda in Andhra Pradesh state (Goel & Dube 1999).

In Maharashtra, beautiful and elaborate rock tunnels, rock temples are famous for underground engineering. They were made out of the hardest rock and the tunnels were built for some kilometers. The tunnels of Ellora alone add up to 10.8 km in length. In medieval India, forts and palaces were provided with fountains, underground pathways and basement halls for storage, meeting halls, summer retreats and water tunnels. The underground constructions in Daulatabad fort, Man Mandir in the palace of King Man Singh and the 17 basement chambers below the famous Tajmahal, are outstanding constructions of medieval kings of India (Sharma & Selby 1989).

Step wells, Adalaj, India are underground structures dug deep trenches for water wells reached by stairs or steps called baoli, vav, kunds, bawdi, kalyani or pushkarni. Adalaj is one of the finest found in Gandhi nagar. The step wells are used not only for daily water needs but were considered to be sacred. The place was used to socialize and gather for religious ceremonies and virtually as subterranean temples with exquisite carvings and idols of male and female deities(Shuichi Takezawa *et al.* 2002, Sharad Chandra 2015).

The mysterious 8 km tunnel named Qutb Shahi connects Charminar and Golconda fort for the safe escape of the imperial family or for secret messages. The Mughals too were masters in tunnel construction, connecting various places like Delhi, Agra and Lahore. The tunnel of Srirangapatna temple is a noteworthy example found in Tamilnadu is sensational archeological discoveries. The tunnel was used by the members of royal family and their military generals as escape route and also an underground treasure vault (The Hindu).

Hydropower projects in Delhi consist of larger length of tunnels with sizes varying from 2.5 to 14 m diameter to add about 16500MW of hydropower. After the success of metro rail project in Delhi with the state-of the-art technology, construction of metro rail project is being planned in various cities including Mumbai, Bangalore, Hyderabad, Lucknow, Pune, Chandigarh, Jaipur and Howrah-Kolkata. The Indian Railways is constructing the most challenging Jammu-Udhampur-Srinagar Baramulla railway line in the difficult Himalayan terrain of Jammu and Kashmir State and there are more than 40 tunnels of total length 107.96km in the Katra-Quazigund section (142km). The Konkan Railway Corporation Limited (KRCL) has constructed Konkan railway line with 92 tunnels of total length 83.6km. The Border Roads Organization (BRO) is also constructing a challenging Rohtang highway tunnel with a length of 8.9km at an altitude of 3978m on Manali-Leh road. National Highway Authority of India (NHAI) has planned a number of ambitious projects including the widening of NH1A having the longest Chenani Nashri highway tunnel (Goel *et al.* (2016) Underground Infrastructures: Planning, Design, and Construction).

2.2.2 International Development

China is famous for underground developments. In Shangai, China more than 2 million m^2 underground space is utilized for sub surface structures. Underground supermarkets, warehouses, silos, garages, hospitals, markets, restaurants, theaters, hotels, entertainment centers, factories and workshops, culture farms, plantations, subways and subaqueous tunnels may be found throughout the city of Shanghai .One outstanding example of the use of underground space is an underground ice hockey stadium with a span of 61 m, in Gjovik, Norway, built for the 1994 winter Olympic Games. (Xueyuan & Yu 1988).

In 1905, El Teniente mine, the world's largest copper mine, south of Santiago, Chile, started its expansion. It was located in the Andes mountain range, continues its growth by digging ever deeper to extend its productive life. The mine has nearly 2,000 miles of underground tunnels and almost 1,000 miles of underground roads.

The world's largest underground flood diversion facility is named as Metropolitan Area Outer Underground Discharge Channel, Japan. It has 72 feet below the ground about 20 miles from Tokyo, the facility has 59 pillars each weighing 500 tons that support the ceiling over this 580-foot-long, 255-foot-wide structure.

The Gotthard Road Tunnel, Switzerland which is 16.9 kilometers in length is one of the three tunnels that connect the Swiss Plateau to southern Switzerland and run under the Gotthard Massif, the two other being railway tunnels, the Gotthard Tunnel (1882) and the Gotthard Base Tunnel (2016).

Zurich Stadelhofen is an important local railway station in the city of Zurich, on the Zurich-Rapperswil and Zurich-Winterthur. The rail approaches at both ends from Hirschengraben Tunnel to Zurich Hauptbahnhof to the north. To the south the line divides inside the tunnel, with one route traversing the Zurichberg Tunnel to Stettbach station and the other a single track tunnel to Tiefenbrunnen station. The station can be accessed from either side. An underground retail arcade runs the length of the station below the tracks and provides access between the platforms and station entrances. Underground access is supplemented by two bridges which span the station, one carrying a footpath and the other restricted road traffic.

Wieliczka Salt Mine, Poland, 1,000-foot-deep mine that stretches around 180 miles in length. Sanford Lab, South Dakota the continent's largest and deepest former gold mine 5,000 feet below ground. It consists of room for expansion, as the mine dips 8,000 feet below the surface. Stretching to the equivalent of more than six Empire State buildings, the lab at 4,850 feet deep allows the study of dark matter, shielded from interfering cosmic rays by the mass of earth above it.

The Boston Central Artery/Tunnel project also known as the "Big Dig" is one of the largest and most complex urban transportation projects ever ventured in the United States. It provides solution for the major traffic allows flow and snarls and 200,000 vehicles a day through the center of downtown Boston. The Ted Williams Tunnel interface is 90 feet below the surface of Boston Harbor, the deepest such connection in North America. The project's underground utility relocation program moved 29 miles of gas, electric, telephone, sewer, water, and other utility lines maintained by 31 separate companies. 5000 miles of fiber optic cable and 200,000 miles of copper telephone cable were installed (Master Builders).

2.3   ISSUES OF UNDERGROUND DEVELOPMENTS

The Urbanization process creates drastic impacts on catchment hydrology. More than 1 million inhabitants are dwelling in cities. These urban sprawl results in increased run off rates and volumes. Base flow and the losses due to infiltration are the other negative effects of urbanization. Development of infrastructures results in the creation of impervious areas and the drainage work simplification results in faster runoff with reduced recession time and shorter time concentration. The Ground water dynamics is totally getting altered and the ground water regimes inflow and outflow are not balanced.

The barrier created below the ground surface creates obstruction to the flow and causes changes in the head at various points. Due to the changes in the slope the sediments and concentrations carried out by the ground water flow changes the ground water quality due to the deposition and stagnation of constituents.

2.3.1   Impact of Urbanization

The cities located above shallow unconfirmed aquifer experience the most important issues in the interaction between urban development and groundwater, especially in the growing cities. The interaction between urban development and groundwater depend upon the pattern and stage of city evolution on affecting the quantity and quality of groundwater. The increased groundwater abstraction and the existing new sources of recharge are the main cause that brings the changes in quantity and quality of ground water below the ground surface. Foster *et al.* (1994) and Morris *et al.* (1994) studied the effect of urbanized area on groundwater. They identified two main issues namely 1) urbanized area changes groundwater recharge or cycle, with modification to the existing recharge and the introduction of the new sources,

2) the introduction of new sources of recharge in urbanized area causes extensive but essentially diffuse groundwater contamination. Three main problems are arise due to the above mentioned threats related with groundwater under growing cities: (1) fluctuations in groundwater levels, (2) severe groundwater contamination and (3) impact on engineering structure (Vazquez-Sune *et al.* 2005).

The status of groundwater in Chennai city was studied using physicochemical and biological parameters according to the standard methods (APHA 1998). Two zones (North and South) from Chennai city were selected for the study and from each Zone 25 sampling stations were fixed and the analysis was made during summer and monsoon seasons (Jan – Dec) 2007. Results indicated that the groundwater of the study area was bacteriologically not safe and needed treatment before it was used for drinking purposes (Loganathan *et al.* 2011).

Sarfaraz Ahmad *et al.* (2008) evaluated the impact of urbanization on the surface and sub surface hydrological regimes. The urban migration leads to rapid urbanization which pollutes the water tremendously making the water resource allocation very complex. The geographical and geological set-ups of the studies are very important in determining the effect of urbanization to a greater extent.

Syeda Jesmin Haque *et al.* (2013) attempted to realize the challenges that are faced by mega cities and the south asian developing countries due to urban rejuvenation in groundwater conditioned by canal seepage and immense sewage loads. Numerous studies were reviewed, to understand the connection between groundwater pollutants and urbanization. The data obtained from a variety of national and international organizations were analyzed to study the effects of this two interaction. In Delhi and Dhaka, the over extraction of ground water for the development has lowered the groundwater level to the greater extent, particularly the densely populated Delhi experiences the vulnerable groundwater pollution.

This pollution in the coastal areas of Karachi and Mumbai is magnified by the proximity of seawater as well as the increasing populations of these areas. The study examined that chlorides and nitrates are the chief anthropogenic toxins being the major groundwater pollutants associated with this urbanization. This paper also concluded with policy recommendations for minimizing the impact of urbanization on groundwater.

The rapid urbanization has exerted heavy pressure on land and water resources in cities resulting in serious environmental and social problems (Leao *et al.* 2001). In the last 200 years, world population has increased six times, and the urban population has multiplied 100 times (Radzicki 1995).

Urbanization occupies a puzzling position in country development and growth, global economic development, energy consumption, natural resource use, and human well-being (McDonald *et al.* 2011, 2014, Uddameri *et al.* 2014, Mondal *et al.* 2015, Pandey & Joshi 2015, Jain *et al.* 2016).

Underground development is an important strategyin developing and reshaping urban areas to meet the challenges of the future. Placement of infrastructure and other facilities underground presents an opportunity for realizing new functions in urban areas without destroying heritages or negatively impacting the surface environment (Wout Broere 2016).

Infrastructure developments such as tunneled metro rail corridor often interfere directly with water resources. The changes due to the subsurface structure cause numerous anthropogenic impacts make urban geological and hydro geological issues complex (Huggenberger & Epting 2011).

Tunneling beneath the groundwater table causes changes in the state of stress and the pore water pressure distribution. In such tunneling problems, the tunneling work inevitably causes water inflows into excavated area, thus causing the change in the pore water pressure distribution. Thedirect environmental consequence of water inflows during tunneling is the drawdown of groundwater level in the surrounding aquifer. The related ground subsidence occurring as a result of the reduction in water pressures in the soil layers can damage nearby structures/utilities (Yoo 2005).

During tunneling the trenches are excavated and to keep it dry groundwater should be pumped out from its bottom part. This causes lowering of groundwater table in the surroundings of the trench (Aivars Spalvins *et al.* 2012). Due to tunneling the potential to cause groundwater drawdown and the ground settlement that may affect existing structures will be an issue (Sian France *et al.* 2010).

In the southern region of U.S urbanization creates a major impact in the changes in the streams. The channel disturbances along streams and increased potential for hydrological alterations are the major impacts of urbanization.

Because of differences in climatic inputs, vegetation, geology, slope, stream geomorphology, and hydrologic processes the urban stream responses may differ across physiographic regions. To improve the stream response to land use change, there is a growing need to synthesize regional hydrologic responses to urbanization. In the southern U.S. the watershed management efforts guide the future research in the rapidly urbanizing region.

Foster *et al.* (1998) studied the urbanization effects on quantity and quality of the underlying groundwater. The changing patterns of rates of recharge and initiating new abstraction regimes adversely affect the groundwater quality.

The construction of roads, buildings and car parks affects routes of infiltration by any change that makes the land surface more im-permeable. Recharge patterns also get modified in the natural sources. The changes in natural drainage by canalisation of streams, construction of storm water drains and soak away will collect the rainwater from these impermeable surfaces and produce locally-concentrated infiltration (Lerner 2002; 2004). The leakage water from main and sewerage networks beneath the ground water large volumes of additional infiltration. As cities become larger, the water infrastructure may increasingly be dependent on surface water or groundwater brought in from outside the urban area itself. Other potential sources of additional urban recharge include on-site sanitation systems and the irrigation of amenity areas such as parks and sports grounds (Morris *et al.* 2003).

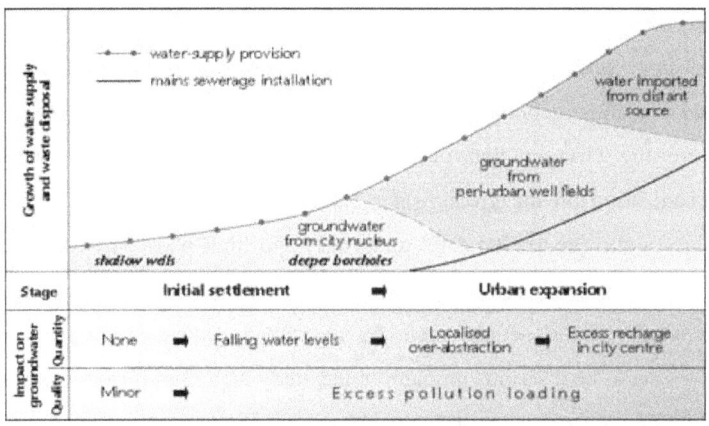

(Source : Morris *et al.* 2003)

Figure 2.1 Evolution of urban water infrastructure

Khazaei *et al.* (2004) investigated Zahedan aquifer for the impacts of urban growth on groundwater quality and quantity, which is the sole source of water supply for the city of Zahedan, Iran. During December 2000, groundwater levels in 40 wells were selected as observation wells. The investigation has based on the collection of available historical data, supplemented by field and laboratory investigations. And during November and December 2000, 102 water samples were taken in addition to the existing data. Of these, 43 samples were analyzed for major ions, 32 samples were analyzed for nitrogen and phosphorus and the remainder for bacteriological contamination. The water level data seemed to be declined due to over-abstraction since 1977. The magnitude of this decline reached about 20 m in some places. However, in one area over the same period, a rise of about 3 m was observed.

The local hydro geological conditions of shallow bedrock and relatively low permeability materials downstream of this area that limits the flow of groundwater towards the northeastern part of the aquifer. The general fall in groundwater levels, changes in the direction of the groundwater flow and overall reduction in the areal extent of the saturated region of the aquifer were the changes that occurred in the aquifer. The heavy impact of urbanization on the groundwater quality was shown through the observed high nitrate (up to 295 mg/l as nitrate) and high phosphorus values (about 0.1 mg/l as P). As the water would not be sufficient enough to meet the demands of the people even the abstracted ground water was rationed. It seems to be a serious problem in the study area. Significant changes in the chloride concentration are also observed in two areas: increasing from 100 mg/l to 1,600 mg/l and from 2,000 mg/l to 4,000 mg/l, respectively.

Bacteriological investigations showed that 33 percent of the 27 collected groundwater samples were positive for total coliform and 11 percent of the samples contained fecal coliforms indicating that local sources were strongly influencing the observed chemical data. Greater depths to groundwater reduced the observation of coliform contamination. In general, the unplanned urban development in Zahedan had significantly degraded the region's water resources and significant actions such as upgrading the sewage waste disposal system, locating other sources of water supply, and strict groundwater management would all be needed to resolve the problems that had arisen.

2.3.2   Head Declination

Jamshidzadeh & Mirbagheri (2010) calculated the mean water table level depletion between a period of time and the mean depletion rate of water table. In this study 21 sampling wells and 53 observation wells were analyzed to evaluate groundwater quality and quantity. According to these data, the mean water table had decreased from 871.75 m in 1990 to 863.82 m in 2006, indicating a mean water table decline of 0.496 m/year. The rate of water withdrawal has increased from 70million cubic meters in1965 to 239 million cubic meters in 2003, indicating an average exploitation ratio of more than 340%. The physicochemical characteristics of groundwater such as pH, hardness, chloride (Cl) and Electrical Conductivity (EC) and Total Dissolved Solid (TDS) values were also calculated.

Marufur Rahman & Mahbub (2012) calculated the change of groundwater level with expansion of irrigation in Bangladesh. Secondary data was mainly used for this study. Hydrograph analysis, groundwater level mapping, groundwater depletion rate calculation were done from groundwater level observation well data of Bangladesh Water Development Board (BWDB).

Mapping software Arc GIS 9.3.1 was used for mapping. As a result the difference between maximum and minimum water level in one season was 2.67 ft. The average value of yearly maximum rate of depletion and minimum rate of depletion was 1.04 ft/year.

Youssef *et al.* (2012) assessed the Groundwater Resources Management in Wadi El-Farigh Area Using MODFLOW. The results of the model threatened the sustainability of the development in the area. The maximum groundwater decline applying the current exploitation strategy (303703 m3/day)would reach 30m after 7 years while the decline would reach 35m in case of increasing the pumping rate by 15%. In addition, construction of the proposed new irrigation canal in the NE direction of study area will improve the groundwater recharge (maximum groundwater decline of 16m). To conserve the storage for longer time, it was recommended to reduce the number of the pumping wells (not more than 800 wells), reduce the initial and running time (not more than 12 hours), applying discrete irrigation system and achieving the objective of implementing the development policy with the groundwater recharge from the proposed new canal.

2.3.3   Sea Water Intrusion

Foster *et al.* (1998) reported that the groundwater resources were rarely sufficient to satisfy the water demand of the larger cities and the issue of resource sustainability arose. Aquifer depletion induces contamination of seepage water and the intrusion of coastal saline water in unconfined aquifers. The serious consequences of urban infrastructure result in land subsidence and the settlement of inter-bedded aquitards in semi-confined aquifers.

Awanindra Pratap Singh *et al.* (2013) predicted the sea water intrusion in the mining lease area of Surka [District Bhavnagar, Gujarat (India)]. It is located in Surka, Bhavnagar, and Gujarat (India) 6–12 km horizontal distance in the sea shore of Gulf of Cambay. They investigated the whole mining block to determine the relationship between fresh and saline water using Ghyben-Herzberg relation. They also did the Electrical Resistivity Survey and groundwater table monitoring along with ground truth verification with the help of remote sensing in 2004. Due to the close proximity of the sea shore the mining lease area experienced the intrusion of sea water threat during the onset of lignite extraction. The result showed that there was no sea water intrusion as per the Ghyben-Herzberg relation. But if there would be excess pumping of water the up coning of saline water interface would become a hindrance.

Mathurin *et al.* (2012) studied the Effect of Tunnel Excavation on Source and Mixing of Groundwater in a Coastal Granitoidic Fracture Network. The study aimed to assessing the impact of Aspo Hard Rock Laboratory tunnel in the pollution of ground water in fractured crystalline (granitoidic) bedrock. The cross section of the tunnel was 3600 m long and e depth of 460 m. It located at a coastal site in Boreal Europe. Their study consisted of 1117 observations from 180 samples collected from various ground water samples. A classification system was developed to relate the groundwater observations to source and post infiltration mixing phenomena on the basis of the values.

The results showed that the groundwater had multiple sources and a complex history of transport and mixing, and was composed of at least glacial water, marine water, recent meteoric water, and old saline water. The larger impact on flow, sources, and mixing of the groundwater upflow of deep-lying saline water, extensive intrusion of current Baltic Sea water, and substantial temporal variability of chloride were found in many fractures.

2.3.4   Water Quality Deterioration

Foster *et al.* (1999) explained that, due to the high proportion of urban population the link between groundwater and sanitation seems to be very important in developing nations. If the ground water and sanitation is not good, it results in substantial subsurface contamination and a hazard to groundwater quality .Most of the aquifers have sufficient natural groundwater protection to eliminate faecal pathogens in percolating wastewater from in-situ sanitation. The sub-standard water well construction and/or informal or illegal sanitation and waste disposal practices are likely to increase the hazards (Lewis *et al.* 1980).

Nitrogen compounds usually nitrate and dissolved organic carbon resulting from pharmaceuticals, disinfectants, and detergents found to be in a greater depth and persist after the contamination source is removed are found to be troublesome (Foster *et al.* 2010).

Different parameters such as pH, turbidity, Alkalinity, total hardness, chloride, sulphate, nitrate, TDS, DO, BOD, COD were evaluated using water quality index for ground water of Bapatla mandal, coastal Andhra Pradesh in India. The TDS, nitrate and calcium levels in all the samples exceeded the Maximum Contaminant Level (MCL) of WHO (World Health Organization), BIS (Bureau of Indian Standards) and USPH (U.S. public health) Drinking Water Standards (Sudhakar Gummadi *et al.* 2014).

Asadi *et al.* (2008) studied the water quality index of groundwater for Khandaleru catchments, Nellore district, Andhra Pradesh. The attribute database was created from the groundwater samples collected at the predetermined locations. The samples were analyzed for physico-chemical parameters and used for the generation of attribute database. The spatial distribution maps were created for the water quality parameters such as pH, alkalinity, chlorides, sulphates, nitrates, hardness, TDS, fluorides and sodium. The results of the analysis maps were used for curve-fitting method in GIS software. The computation of WQI and the analysis of the physico-chemical properties are helpful in the grouping of groundwater samples into excellent, good, poor, very poor, and unfit.

Analysis of water quality parameters of groundwater near Ambattur industrial area Tamilnadu, India was carried out with regard to the parameters such as pH, total alkalinity, total hardness, turbidity, chloride, sulphate, fluoride, total dissolved solids and conductivity. Comparison of the physio-chemical parameters of the water sample with World Health Organization (WHO) and Indian Council of Medical Research (ICMR) limits showed that the groundwater is highly contaminated and accounted for the health hazards making it unfit for human use (Saravanakumar *et al.* 2011).

Foster (1990) reported that in Latin America and the Caribbean, Industrial development and urbanization jointly created profound impacts on the hydrological cycle. The effects such as increased magnitude of peak flood runoff and changes in the surface water flow caused deterioration in river water quality. Changes in groundwater recharge regime also often occurred as an adverse impact in the water table condition. The type and the stage of urbanization determine the magnitude of the effects caused by the changes in the ground water regime. This work identified and analyzed the urban impacts on groundwater. These unanticipated collateral environmental and health effects results due to the inefficient water resource planning. For the future prediction and control, efficient water resources management is unavoidable.

Throughout the world the rapid expansion of major cities becomes the cause for the degradation of water quality in local aquifers. In the middle of the 20th century, due to the increased use of road deicers the cities in Canada, northern United States and Europe ground water quality seems to have been degraded. Similarly in the historical context the cities like Chicago, Illinois, and its outlying suburban area can be shown as an example of the effects of urbanization.

Kelly (2008) has done a statistical study of historical water quality data to determine how urbanization activities have affected ground water quality particularly in the shallow (<60m) water. The counties west and south of Chicago experienced the increased Chloride (Cl-) concentrations .During the 1960s the western and southern counties showed that the majority of shallow public supply wells had increased Chloride (Cl-) concentrations. The samples collected from the Chicago area public supply wells had Cl-concentrations greater than 100 mg/L and the medium values were less than10 mg/L.About 43% of the wells in these counties had rate increases greater than 1 mg/L/year, and 15 % had increases greater than 4mg/L/year. Due to both natural and anthropogenic factors, the Cl- concentrations showed a greater increase in the outer counties. These increased concentrations were found to have occurred predominantly in the shallower sand and gravel deposits and in the curbing of major highways and streets, and less development in some parts of these counties.

Foster *et al.* (1994) stated that 'groundwater cycle' got modified greatly due to urbanization with some benefits but with greater threats. The impacts to the ground water regime were mainly due to the Urbanization and its associated industrialization in the study area. The man–made modifications to the ground water regime create a marked impact on aquifers underlying cities. These changes create serious impacts on urban infrastructure (Foster *et al.* 1998; Howard 2007). These result in the modifications in the systematic variation and the hydrogeological settings. The aquifers that allow free vertical movement of water are called unconfined (oxygenated) aquifers and there is a direct interaction between the pollutants to the water-table and the built infrastructure. The vertical water movement was greatly impeded by the confined aquifers often containing anoxic groundwater. They are less prone to pollution but more readily over-exploited.

2.4   TOOLS AND TECHNIQUES

The various tools available for graphical interpretation and representation are remote sensing techniques and modeling techniques. Geographic Information System (GIS) uses various methodologies to interpret various outputs for the given inputs. Modeling techniques such as Visual MODFLOW are used to develop a conceptual model to predict the changes in the future.

2.4.1   Ground Water Modeling

Gregorauskas *et al.* (1999) carried out a model on groundwater flow and contaminant transport at Klaipeda oil terminal, Lithuania and found that water table aquifer in the area of Klaipeda oil terminal was polluted with hydrocarbons (oil products) dissolved in water. Above this aquifer, there was a layer containing oil and reaching 0.5 m in thickness. These pollutants did not threaten drinking water sources, but oil could enter the lagoon of Kursiu Marios and the Baltic Sea. Groundwater monitoring was organized: shallow groundwater investigations were done; filtration and migration models of the terminal and adjacent areas were consulted. Modeling results showed that the flow of hydrocarbons to the lagoon could be efficiently barred by a horizontal drain.

Groundwater models describe the groundwater flow and transport processes using mathematical equations based on certain simplifying assumptions. These assumptions typically involve the direction of flow, geometry of the aquifer, the heterogeneity or anisotropy of sediments or bedrock within the aquifer, the contaminant transport mechanisms and chemical reactions (Kumar 2000).

Barathi (2004) carried out Groundwater quality modeling of Chennai Nandanam area. She simulated groundwater flow using Visual MODFLOW. The aquifer characteristics, water level data for the observation wells and rainfall data were used to run the model. The model was run to simulate water level in 2001 and validated with field data by comparing with the hydraulic head of observation wells. Other outputs such as ground water flow and movement were obtained for the period of 2001. The volumetric budget for the model was checked and the changes in storage of groundwater system were assessed. The calibrated model was used in determining the quantity of recharge and the optional location for recharge to improve the groundwater storage. The author concluded that there was a good agreement between the computed values and filed data.

The author has studied the water level in Adyar river. Adyar river acts as a drainage during November 2001 and March 2004 (monsoon and post monsoon season), and the only source for recharge is the rainfall. The velocity increases as the flow moves towards the river. The velocity of flow is high in the central part and also in the north-east and south-west part indicating recharging in these areas may drain more quickly into the river and canals. The total inflow into and outflow from the system are nearly equal with a percentage discrepancy of 0.03 which indicates proper execution of the model simulation. The author also recommended that modeling of salt water intrusion into the coastal aquifer using Visual MODFLOW.

Visual MODFLOW software package includes three main software and many support modules. MODFLOW software is used to calculate the volume, quality and distribution of groundwater flows. Function of MODPATH software is calculating the direction and speed of flow when it moves through aquifer system. MT3D software is used to calculate diffusion and transportation processes with chemical reaction of solutes in groundwater flow system (Lakshmi priya *et al.* 2015).

MODFLOW is a groundwater flow model that numerically solves groundwater flows for a porous medium (Harbaugh *et al.* 2000). In MODFLOW, an aquifer system is divided into rectangular blocks on a grid, which is organized into rows and columns .Each grid represents a layer, with several layers stacked upon each other to represent different soils and layers in the aquifer in three dimensions. Layers can be defined as confined, unconfined, or a combination of both. Groundwater flow is determined from parameters such as hydraulic conductivity, water table depths, ET, recharge rate, and depth of each layer (Paul M Barlow *et al.* 2006).

Hassanizadeh & Leijinse (1988) worked on the modeling of brine transport in porous media. They discussed certain important physical and mathematical differences between low and high concentrations situations. They solved a set of two non-linear coupled partial differential equations obtained from a modified formulation of Darcy's and Fick's laws by means of iterative methods. Lassey (1988) derived an analytical solution to the advection-dispersion equation for one-dimensional solute or tracer transport including sorption and first-order loss.

Miller & Weber (1988) described laboratory investigations and mathematical modeling of the sorption of hydrophobic solutes by aquifer materials. They obtained accurate representations of the sorption process with either a dual resistance diffusion model or an equilibrium/first-order sorption rate model.

Measured water levels matched simulated water levels under varying meteorological (precipitation, ET, and recharge) conditions. Error between measured and simulated groundwater levels was below 2% (Bradley *et al.* 2000). Hydraulic conductivity (K) is the "measure of the ability of fractured or porous media to transmit water" (Fetter 1999). Higher K values represent media through which water may pass easily, and media with low K values are more impermeable to water flow (Freeze & Cherry 1979).

John Doherty (2001) reported that use of USGS ground water flow model MODFLOW is often hampered by the occurrence of 'dry cells'. While MODFLOW allows such cells to 'rewet' in the course of a simulation, stability of the heads solution process is often problematic with rewetting functionality operative. In many cases of practical interest, particularly in mining applications, MODFLOW simply fails to converge. However, by making a number of adjustments to the MODFLOW block-centered flow package, it is possible to overcome this problem in many instances of MODFLOW development. These adjustments are such as to allow a layer to transmit water, albeit with vastly reduced transmissivity, even if the water level in that layer is below its base. With these alterations MODFLOW cells can remain active even if they lie within the unsaturated zone.

Anandha Kumar & Sinha (2003) undertook the Groundwater simulation study for the Hirakud command area. A mathematical model had been developed using MODFLOW package of USGS to simulate hydrogeological condition of the groundwater flow systems in the command area .The main objective of the study was to arrest the ever rising ground water levels as well as to control further extension of water logged areas. The groundwater simulation studies have showed that the water logging condition prevailing in part of the Hirakud command area could be controlled by the development of groundwater in conjunction with surface water without any deterioration to the groundwater regime.

Haitjema (1992) performed a ground-water flow modeling study for the Four County Landfill (FCL) in Fulton County, Indiana. The modeling was performed with the relatively new analytic element method. The new method employs superposition of closed form analytic solutions, rather than a grid or element network, and proved an effective tool in answering some unresolved hydrogeological questions. The study results demonstrated substantially higher regional and local hydraulic conductivities than that suggested by the landfill consultants, well-connected upper and lower aquifer zones at the site, and potential ground-water travel times from the landfill to the nearby Tippecanoe River of less than 15 years.

2.4.2   Remote Sensing/GIS

Various thematic layers such as base map, drainage map, counter map, geology map were prepared for the entire basin of river tons in Allahabad district, Uttar Pradesh, India using GIS and remote sensing techniques using Arc GIS 9.3 software. These thematic layers had been integrated on GIS environment through assigning proper weight to various factors controlling occurrence of groundwater. As a result, groundwater potential zones map obtained, classified the study area into zones such as very good, good, moderate, poor and very poor (Shukla 2014).

Abbas Almasi *et al.* (2014) interpolated the spatial analysis that used in GIS which is applied for soil analysis. With Inverse Distance Weighted (IDW) one can control the significance of known points on the interpolated values based on their distance from the output point. This could be used to determine the distribution of a continuous data, phenomenon like rainfall, temperature, humidity, soil or distribution soil properties. He found both kriging and IDW methodology would be useful in analyzing soil parameters.

Asadi *et al.* (2007) applied remote sensing and GIS techniques for evaluation of groundwater quality in Municipal Corporation of Hyderabad (Zone-V). The Groundwater quality in Hyderabad has special significance and needs great attention of all concerned since it is the major alternative source of domestic, industrial and drinking water supply. The study monitored the groundwater quality, relates it to the land use / land cover and maps such quality using remote sensing and GIS techniques for a part of Hyderabad metropolis. Thematic maps for the study were prepared by visual interpretation of SOI topo sheets and linearly enhanced fused data of IRS-ID PAN and LISSIII imagery on 1:50,000 scale using AutoCAD and ARC/INFO software. Water Quality Index (WQI) was then calculated to find the suitability of water for drinking purpose. The overall view of the water quality index of the present study area revealed that most of the study area with > 50 standard rating of water quality index exhibited poor, very poor and unfit water quality except in places like Banjara Hills, Erragadda and Tolichowki. Appropriate methods for improving the water quality in affected areas had been suggested.

Groundwater Potential Index (GWPI) map of the Araniar river basin was done through an overlay analysis of climatic, geologic, geomorphic, soil and land use/land cover features of the basin using Landsat 5 Thematic Mapper (TM) data and ArcGIS 9.2. Correlation analysis was carried out for rainfall, geology, soil, slope, geomorphology and land use/land cover maps being overlaid with standardized weights of 0.49, 0.20, 0.17, 0.05, 0.05, and 0.04 and maximum correlation coefficient of 0.922 was obtained. The GWPI map showed groundwater potential zones as "excellent", "very good", "good", "moderate" and "poor" with yield values in the ranges 293–361, 210–292, 126–209, 43–125, and 15–42 lpm, respectively (Jasmine *et al.* 2014).

Using Secondary data, changes in groundwater level with expansion of irrigation in Bangladesh was obtained through Hydrograph analysis, groundwater level mapping and groundwater depletion rate calculation. Mapping software Arc GIS 9.3.1 was used for mapping. The average value of yearly maximum rate of depletion and minimum rate of depletion was seen as 1.04 feet / year (Marufur Rahman *et al.* 2012).

Sui *et al.* (1999) reviewed the practices and problems in the hydrological modeling based on Geographic Information Systems (GIS). The authors argued that integrating GIS with hydrological modeling was very essential. It is a current stand-alone and various loose/tight coupling approach. It is a technology-driven without adequately addressing the conceptual problems involved in the integration. The conceptualizations of space and time embedded in the current generation of GIS are not conceptually compatible with those in the hydrological models. The constraints on the type of hydrological models can be developed. The future research agenda was reframed from the emerging geographic information science perspective.

Estimation of quantitative and qualitative impacts of groundwater due to urbanization was found for Ajmeer, a major city of Rajasthan. Groundwater recharge was computed using the water level fluctuation method. Database related to urbanization and groundwater was created in GIS and the temporal and spatial variations in groundwater quality and quantity were correlated with urban growth, using overlay analysis GIS. Using GIS the average Groundwater recharge was found to be 3.06% (Jat *et al.* 2009).

Interpolated Distance Weighted (IDW) interpolation method and the Water Quality Index (WQI) using Weighted Arithmetic Index method were used to analyze the water quality of Bhadravathi Taluk, Karnataka, India Global Positioning System (GPS) survey, was done. During this analysis, Heavy rainfall and dilution factors, excessive hardness was noticed. The GIS software was employed for interpretation of data. Finally, an Interpolated Distance Weighted map was drawn for (a) pH, (b) Total alkalinity, (c) Total hardness, (d) Calcium, (e) Magnesium, (f) Chloride, (g) Fluoride, (h) Iron, (i) Sodium, (j) Potassium, (k) TDS, (l) Electrical Conductivity (EC) in a different season (Raj Kumar *et al.* 2012).

The physiochemical parameters such as pH, Total Dissolved Solids, Total Hardness, Calcium, Magnesium, Sulphates, Chlorides, Fluorides, and

Nitrates were calculated using Water Quality Index (WQI) was done in and around Ranipet area, Vellore District, Tamilnadu. Thirty five groundwater samples were collected from open and tube wells during the Monsoon and Post Monsoon season in the year 2012. Higher concentration of dissolved solids during post monsoon samples exhibited poor quality of water as compared to Monsoon season due to more seepage and movement of ground water during post-monsoon (Ambiga *et al.* 2013).

Assessment of groundwater quality index was carried out for various parameters such as pH, Electrical Conductivity, Total Dissolved Solids, Alkalinity, Chlorides, Total Hardness, Calcium Hardness, Nitrates, Sulphates, Iron and Fluorides in Bommasandra area, Bangalore city, Karnataka state, India. From the WQI of the water of Bommasandra was considered as poor water (Shiva Prasad *et al.* 2015).

Systematic and comparative evaluation of the water quality was assessed using Water Quality Index (WQI) in Nellikkuzhy Panchayat of Kerala for the water quality parameters. In this result, the seasonal values of WQI indicated that in monsoon ground water was more affected than during summer. This method was more helpful for the public to understand the quality of water as well as being a useful tool in many ways in the field of water quality management (Jai M Paul *et al.* 2014).

Groundwater quality index in vidyanagar, Davanagere city, Karnataka state, India was overviewed for the groundwater quality parameters such as pH, Total hardness, Calcium, Magnesium, Chloride, Nitrate, Sulphate, Electrical conductivity, Total dissolved solids, Iron, Fluoride and Total alkalinity. The results were analyzed by WQI method for predicting water quality (Kalpana *et al.* 2014).

2.5   REMEDIAL MEASURES

Jordi Font-Capo (2012) investigated interaction between groundwater and TBM (Tunnel Boring Machine). He proposed three methodologies to find the interaction between the above mentioned two parameters. The excavated tunnels by the TBM were very sensitive to the sudden changes of the geological media.

1) First, he proposed a methodology that characterized hydrogeologically the medium crossed by the TBM. A quasi-3D numerical model was constructed and different scenarios were calibrated. It is a hydrogeological conceptual model that calibrated the ground water flow. The barrier behavior and the conduit behavior in dikes were the most marked effect and more prominent in faults.

2) Secondly, he proposed a methodology to locate and quantify the inflows in the tunnel face of the TBM being adopted. A major problem encountered was the unexpected high water inflows that resulted in the collapse of the tunnel face and affected surface structures.

3) Third, the hydrogeological impacts caused by tunneling with TBM were characterized. The lining in tunnels reduced water seepage but could cause a barrier effect because of aquifer obstruction. Analytical methods were employed to calculate the gradient and permeability variation after tunneling. The uses of pumping tests allowed determine the barrier effect and the changes in groundwater connectivity due to tunneling.These scenarios enabled them for better understanding to find the correct solutions and to minimize the consequences of tunnel-groundwater interaction.

Vahab Besharat (2012) studied the methods of remediation of the existing underground structures against liquefaction. The methods against liquefaction were classified into following categories: First, to improve the liquefiable soil and prevent liquefaction. The concepts used in the category were: 1: The material that increases the density of existence material should be used. 2- Not-liquefiable grain size should be used. 3- The skeleton of soil should be stabilized. 4- The saturation of soil should be decreased. 5- Immediate dissipation of increased excess pore pressure. 6- Increasing confining pressure by reducing the shear stress 7- To build an underground wall that reduce the shear stress.

Anjana *et al.* (2017) reported that by 2050 the tremendous urbanization would reach upto 70%. When the groundwater level is extremely close to the surface and a linear underground structure is constructed in the ground it blocks the flow of groundwater. To quantify the cumulative impact of underground structures on the flow of urban groundwater was the main aim of this study. For the actual and the potential state of the groundwater flow be assessed with a methodology in an urban area was proposed. All the structures were integrated in the 3D numerical model with their geometry. The modeling process consisted of solving the diffusivity in the porous media equation for the entire area studied. The aquifer regime got modified due to the results showing underground structures fragment groundwater flow systems. Structures with drainage systems showed to have a major impact on flow systems. The comparison between the actual state and the potential state of urban groundwater flow showed that the underground structures had fragmented the flow systems. The following conclusions were drawn from the study:

(1) Fragmentation of urban flow systems was caused by pumping/reinjection rates of the underground structures. Consequently, the urban groundwater regime had been

modified and an inversion of the interaction between the groundwater and the river was observed. This regime modification could lead to an influx of polluted water from the river to the groundwater.

(2) The cumulative effect of underground structures was a global drawdown. Regarding water table elevation and the drawdown was caused by pumping devices (i.e., pumping wells and structures with drainage systems). Due to the strong dependence of the actual water table elevation on pumping devices, the potential state of water table elevation (i.e., water table elevation in case of no pumping device operation) should be taken into account when building new structures. The stoppage of several pumping devices could lead to the flooding of underground floors.

(3) The impact of impervious structures was negligible due to the small hydraulic gradient in the study area. On the other hand, the influence of structures with drainage systems was emphasized. These underground structures had a major influence on urban flow systems.

(4) Transient simulation demonstrated the temporal stability of flow system structure for this case. Thus the relevance of the steady state approach for the quantitative depiction of flow systems was demonstrated. The stability of flow systems was considered an asset regarding underground planning.

2.6   SUMMARY

Assessment of groundwater level changes in the aquifer below the surface level due to tunneling and the water quality analysis have become necessary to analyze because of this artificial barrier below the ground surface. The literature reviewed clearly shows that prediction of these two factors is predominant for the truthful and sustainable urbanization for the future. The literature review has also necessitated the importance of the ground water model through which a perfect model conceptualization is achieved for the study area. Using Ground water model a fruitful and meaningful solution can be arrived as it represents the exact ground truth condition. The model inputs are the key elements which design the model to predict the direction of the flow, water table level changes and the contaminant transportation.

The review still extends that, GIS proves to be widely used graphical tool to represent the scenarios of various periods. It gives the liability of various classifications of different factors in one map. Water quality Index proves to be a worldwide accepted methodology for the water quality analysis and it is a proven tool for finding the intensity of water quality pollution.

A great number of socio - economic issues are rising among the people residing adjacent to the construction. In the literature the socio economic issues are well addressed by the preparation of questionnaire Survey among the public. Different variables included in the questionnaire are used to extract the current scenario about the development. Interlinking of the technical results and Survey output will be useful to frame an effective water management practice in the study area.

CHAPTER 3

STUDY AREA AND DATA COLLECTION

3.1 GENERAL

Chennai is the fourth-largest city in India. According to the 2011 Indian census, it is the sixth-largest city and fourth-most populous urban agglomeration in India. The city together with the adjoining regions constitutes the Chennai Metropolitan Area, which is the 36th-largest urban area by population in the world. To meet the demands of the growing population of Chennai city, urbanization becomes inevitable in their life style. Urbanization again consists of the basic resources like water and energy. Balancing both the future development and reserving the important source like water is very important. Construction of Metro rail Corridors in Chennai has created a major impact in the ground below the Surface as it involves the tunneling techniques. The tunneling below the ground surface creates an enormous change in the underground as it breaks the various heterogeneous layers and the aquifer system.

The tunneling techniques not only make the changes in the water table below the ground surface but also the ground water chemistry. Shied tunneling method was used in the construction and the chemicals like betonite slurry used for piling, pigments and sealants of cement may create changes in the hydro chemistry of the study area. Hence the underground water quality tends to have a change in its nature.

3.2   STUDY AREA

Chennai city is bounded by Kancheepuram district in the south, Tiruvallur district in the north and west and Bay of Bengal in the east. The rivers that flows in the Chennai city are Adyar and Cooum. In Tamil nadu, Chennai is located in the northeastern part. It covers a total area of about 174 Km^2. Underground metro rail corridors of Chennai city were selected as study areas. It is located with the latitude of 13.0827° N, and longitude of 80.2707° E. The Survey of India topo sheet was georeferenced and digitized to delineate the study area around the metro rail corridors. The study area includes the water shed boundaries such as Buckingham canal in the east and Adyar and Coovum rivers in the southern part. The Index map is shown in Figure 3.1.

Chennai is a coastal plain area. The surface and subsurface developments in Chennai lead to the disappearance of water bodies and depressions becomes the major threat for surface and subsurface water resources. Adyar and Buckingham canal serve for the Coastal water sheds. Dense urban watersheds like Saidapet and KK Nagar are covered with river Adyar.Areas like Koyambedu and Aminijikarai are bounded by the river Cooum and they are classified as Less dense sub-urban watersheds.

The surface developments already sealed many of the river sources and again the subsurface developments are affecting the flow path of the ground water. The entire nature of the ground water flow system gets altered and the destruction of the aquifer occurs due to these developments. Hence the study of changes in the ground water regime due to urbanization becomes imperative for the sustainability of the ground water resources.

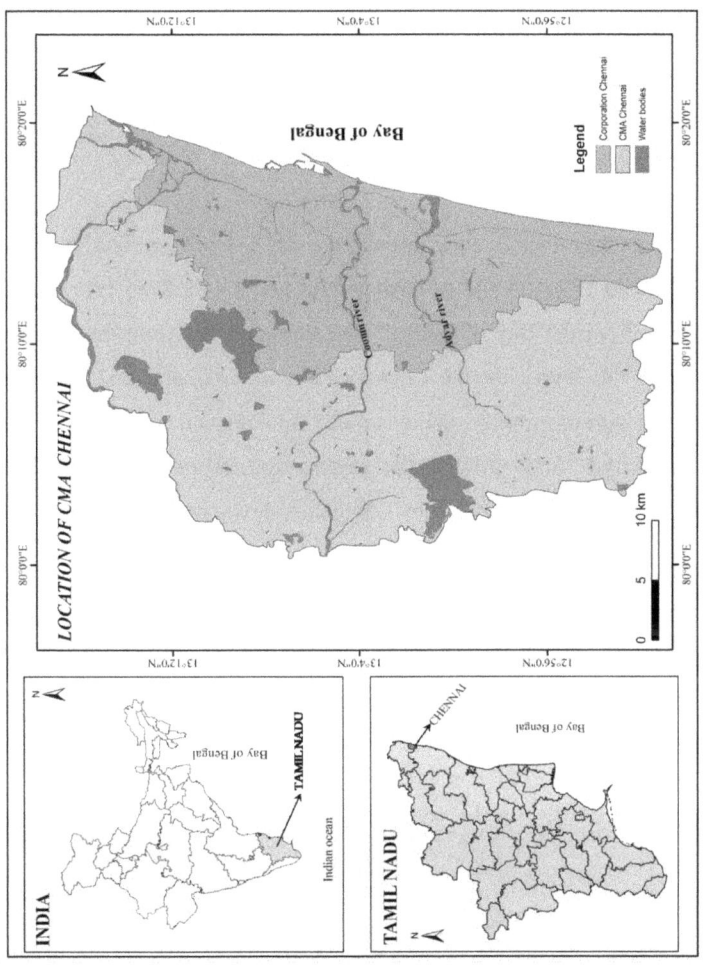

Figure 3.1 Index Map of Study Area

3.3   RIVER BASIN DETAILS

The study area is mainly covered with Adyar and Coovum river basins. Adyar river covers a length of about 12.2 Km and a catchment area of about 33.1 Km^2. It discharges 190 to 940 Mm^3 water annually to the Bay of Bengal. The river gets water only after the Chembarambakkam tank gets filled and the surplus reaches as the orgin flow of Adyar river through the southwest drainage path of Chennai.

The other river flowing through the city is Coovum river. It covers a length of about 17.6 Km and its basin area is about 44.8 Km^2. It originated 70 Km away from the Bay of Bengal. The surplus water from the Coovum tank becomes the source for the Coovum river in Tiruvallur district. Other than the two major river basins the man made Buckingham canal is located in the chennai city.It runs through the eastern coastal strip from north to south.The study area delineated from Chennai is shown in Figure 3.2.

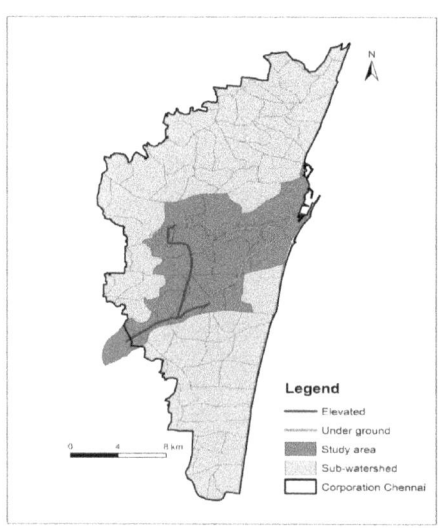

Figure 3.2 Study area delineation from Chennai

3.4   CLIMATE

Chennai is located in the southern part of the country and it is mostly hot and humid. Chennai experiences three major seasons, namely summer, monsoon, and winter. The period from March to June is considered as summer season for Chennai. The season between October and December is referred as the NE monsoon period. The season of Monsoons consists of the months of June to September. The short period from November to February is the winter season for Chennai.

3.5   RAINFALL

Chennai is largely dependent on NE monsoon, since 65% of rains are received in this season. Chennai city receives average moderate rainfall all through the year and is about 1,300 mm. The yearly average rainfall from 1995 - 2017 is shown in Figure 3.3. The highest average yearly rainfall in the city was recorded to be 213.87 mm in 2005.   Recently the city has received the highest average yearly rainfall in the year 2015 due to the massive cyclone occurred in Chennai.

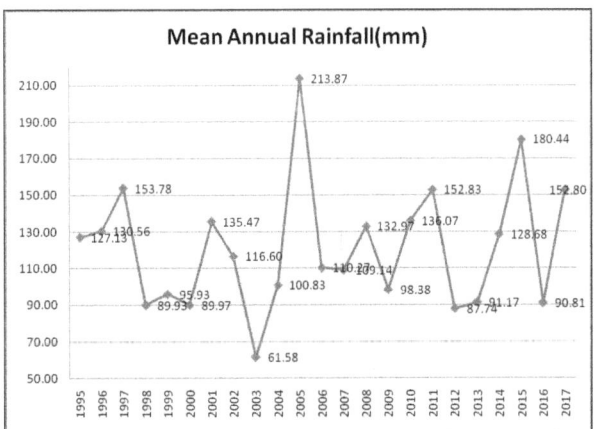

Figure 3.3 Yearly average rainfall in the study area (1995-2017)

3.6   TEMPERATURE

The summer starts around the end of March and continues till June, with late-May to June being the hottest months. During this time, temperatures often cross 40° Celsius. The close proximity to the sea and the thermal equator makes the climate and weather in Chennai relatively consistent with less variation in the seasonal temperature. The average maximum temperature is usually 38° Celsius to 42° Celsius, which is testament of the heat the city experiences during the summer months. The city enjoys a short winter with January being the coolest month when the temperatures dip to around 18 to 20° Celsius.

3.7   HISTORY OF CHENNAI METRO RAIL

The Chennai Metro Rail is a rapid transit system in Chennai, Tamil Nadu, and India. For the rapidly growing population and traffic volumes in Chennai city becomes the need for a new rail based rapid transport system is great. The system provides a fast, reliable, convenient, efficient, modern and economical mode of public transport. The underground corridor covers from Washermenpet to Saidapet with a length of 14.3 kms and the corridor with a length of 9.7 kms. From Chennai Central to Anna Nagar 2nd Avenue will be considered for analysis and it is shown in Figure 3.4.

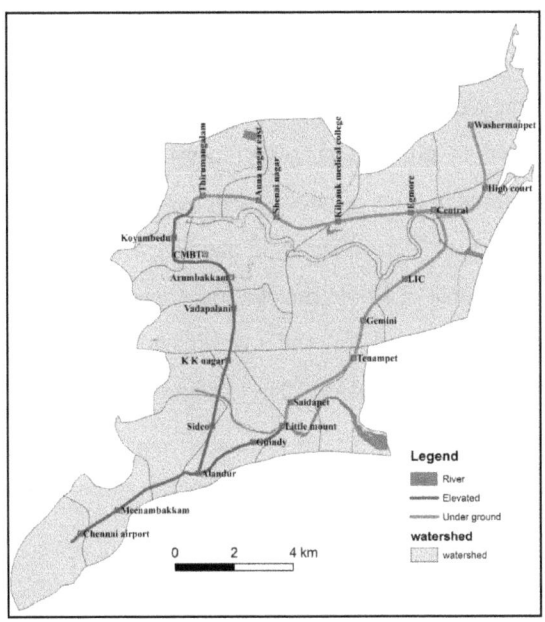

Figure 3.4 Study area with Station Points

Rail level at midsection in tunneling portion shall be kept at least 12.0 m below the ground level so that a cover of 6m is available over the tunnels. At stations, the desirable depth of rail below ground level is 12.5m; Track centre in underground section to be constructed by Tunnel Boring Machine (TBM) is 15.05m to accommodate a 12 m wide island platform. Track centre in underground section to be constructed by cut and cover method is 4.5m.A total of 75 Boreholes were carried out along the corridors in order to study the geology of the area. Provision for environmental impacts of this Metro corridor has been made to cover various protection works, additional compensatory measures, compensation for loss of trees, compensatory aforestation and fencing, monitoring of water quality, air/noise pollution during construction, establishment of Environmental Division.

3.7.1   Progress of Tunneling

The water level changes also depend upon the progress of tunneling work executed in different phases of construction. The various stretches covered distances in different phases such as from January 2012 – December 2013, January 2014 –December 2015 and January 2016 –March 2017. The analysis showed that in all the stretches the completed stretch of tunneling in the earliest phase was very low and it was intensified in the second phase and completed in the third phase. The progress of tunneling is shown in Figure 3.5.

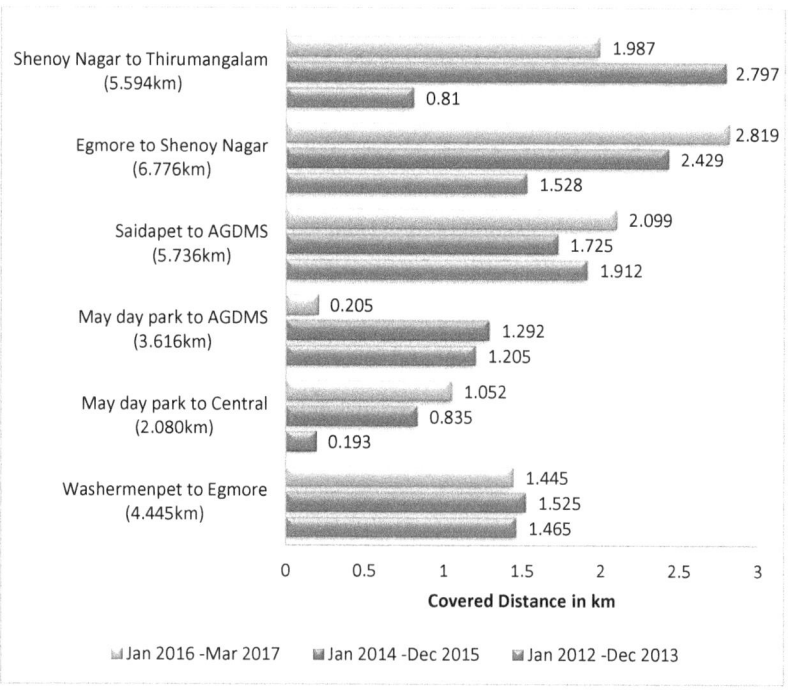

Figure 3.5 Progress of Tunneling

3.8   SOIL

The soil map of the study area was digitized using Arc GIS 9.1 from the soil maps of 1: 50000 scale procured from the Institute of Remote Sensing (IRS), Anna University, and Chennai. Figure 3.6 shows the soil types based on USDA soil taxonomy. The underground stretch that is the eastern part of the study area from Central metro to Saidapet metro is covered with river bank and coasts comprising of fine sand and fine sandy soil where the percolation is very quick. Another or the other underground stretch in the Northern part of the study area, that is from Thirumangalam to Egmore metro is covered with fine silty clay. The lower most southern part of the metro rail corridor is covered with hard rocks.

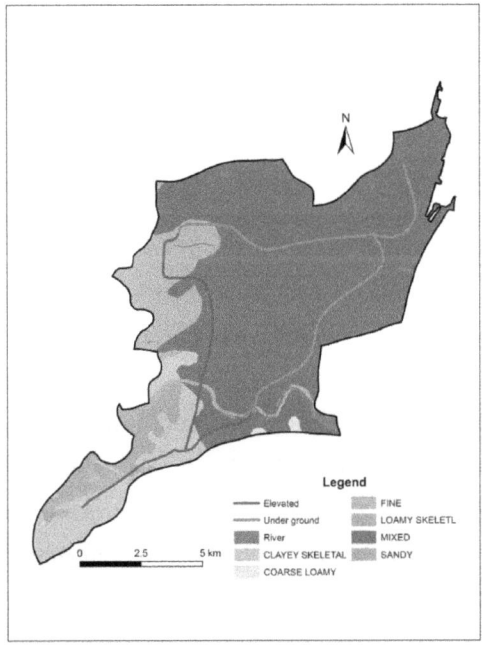

Figure 3.6 Soil Classification

Specific yield is defined as the ratio of the volume of water that a saturated rock or soil will yield by gravity to the total volume of the rock or soft. Specific yield is usually expressed as a percentage. Specific yield values for the various types of soils were identified and assigned to the respective polygons. Table 3.1 lists out the type of soil and its specific yield values.

Table 3.1 Soil Type and their Specific Yield

S.No	Soil Types	Specific yield (%)
1	Coarse loamy	22%
2	Alluvial	26%
3	Fine Sandy	21%
4	Loamy mixed	19.6%
5	Fine loamy	16.4%
6	Fine Silty Clay	18%
7	Loamy skeletal	11.2%

Source: Report of the ground water resource estimation committee, Ministry of water resources, Government of India.

3.9   GEOLOGY

The geology map of the study area was digitized from the map of scale 1:50000 collected from the Geological Survey of India, Chennai. The various geological formations of the study area are shown in Figure 3.7. Chennai district forms part of Tamilnadu coastal plains. Major part of the district is having flat topography with a very gentle slope towards east. Chennai is underlain by various geological formations from the ancient Archeans to the recent alluvium .The geological formations of the district can be grouped into three units: (1) Archean crystalline rocks, (2) consolidated Gondwana and Tertiary sediments, and (3) recent alluvium. The study site consists of coastal sediments in the bank side and a part of archean rocks which carry the composition of Chamockites, Granite and Gneisses as well as fossiliferous rocks intercalated in some places with marine beds of Neocomian age and earliest marine transgression beds of middle Cretaceous ofUpperAlbianage.The other underground stretch from Thirumangalam to Egmore Metro comprises of younger alluvium of varying thickness place to place from 3.00 m to 30.00 m and tertiary of Eocene to Pliocene age which carries the composition of sandstone.

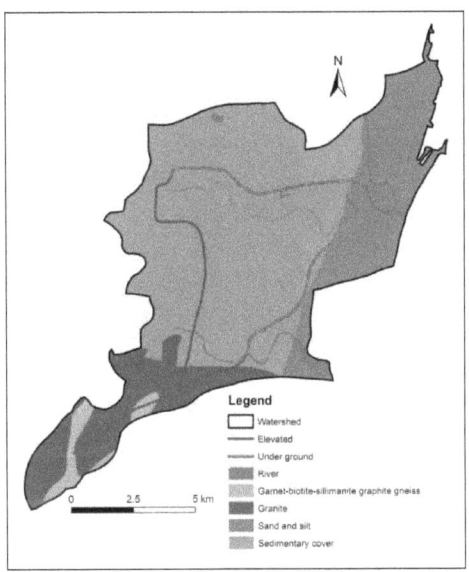

Figure 3.7 Geological classifications of Study area

3.10   GEOMORPHOLOGY

The geomorphology map of the study area has been prepared from the satellite imagery (LISS IV, 5.8 m resolution, September 2018) obtained from Institute of Remote Sensing, Anna University, Chennai. As shown in Figure 3.8, the study area covers from North to South along the Sea shore mainly comprising of Marine deposits. Coastal areas such as beach, beach ridge and deep coastal plain are the various geomorphic units and the plain lands are moderately weathered. The corridor covering from Central to Anna Nagar is mostly covered with fluvial deposits, the deposits which are associated with rivers and streams and the landforms created by them and a part of the extent from North to South along the tunneled area is covered with charconite.

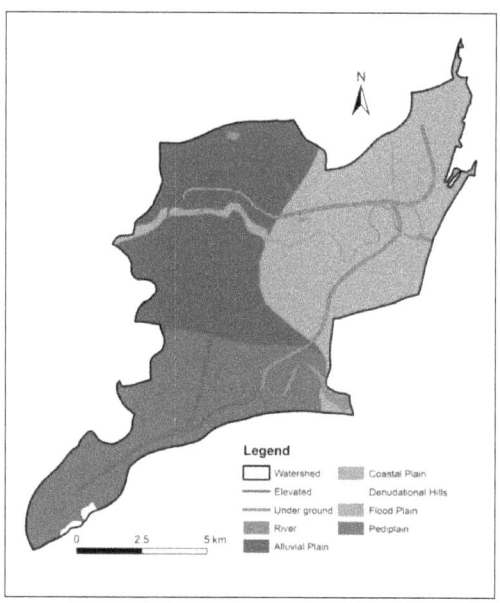

Figure 3.8 Geomorphology map

3.11   HYDROGEOLOGY

Hydrogeology is the area of geology that deals with the distribution and movement of groundwater in the soil and the rocks of the Earth's crust commonly in aquifers.

Bore well datas were collected for the two underground corridor sections from the households dwelling in the nearby areas of the corridor section. The underground stretch from Chennai Central to Anna Nagar covering a length of 9.7 kms mostly comprises of clayey sand, clay and some crystalline rocks. The next underline corridor from Washermenpet to Saidapet covering 14.3 Kms is mostly covered with sand and the coastal sediments like clay and shale.

3.12   SLOPE

Chennai district forms part of coastal plains of Tamil Nadu. Major part of the district is having flat topography with very gentle slope towards east. The altitudes of land surface vary from 10 m above MSL in the west to sea level in the east. Fluvial, marine and erosion landforms are noticed in the district. The statement was proved by mapping the elevation levels of the selected locations of the area and it is shown in Figure 3.9 and the elevation values are tabulated in Table 3.2.

Table 3.2 Locations with Elevation

S.No	Location	Elevation
1	Tondiarpet	12
2	Vepery	12
3	Chepauk	13
4	Lights	13
5	Saidapet	15
7	Aminjikarai	15
8	Tirumangalam	15
9	Vadapalani	18.5
10	K.K.Nagar	16

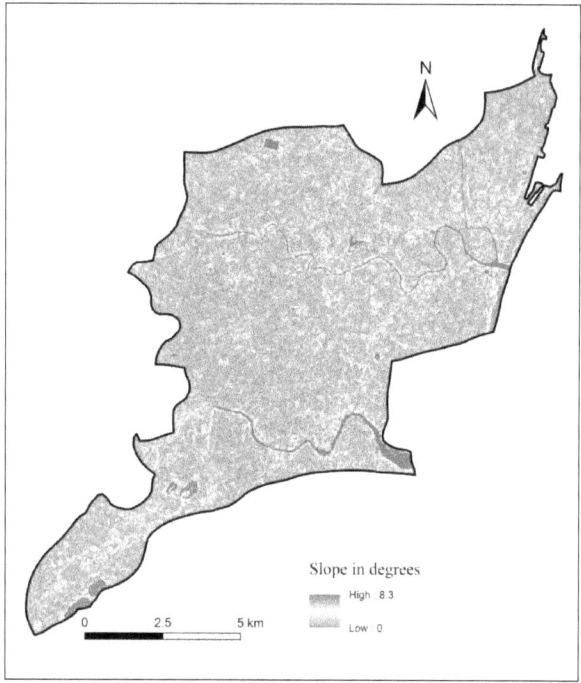

Figure 3.9 Slope classifications

3.13   LANDUSE

The land use map of the study area prepared from the satellite imagery (LISS IV, 5.8 m resolution, September 2018) obtained from Institute of Remote Sensing, Anna University, Chennai, is shown in Figure 3.10. Some of the important factors, which control the land use pattern, are landform, slope, shape of the land, soil and natural resources.

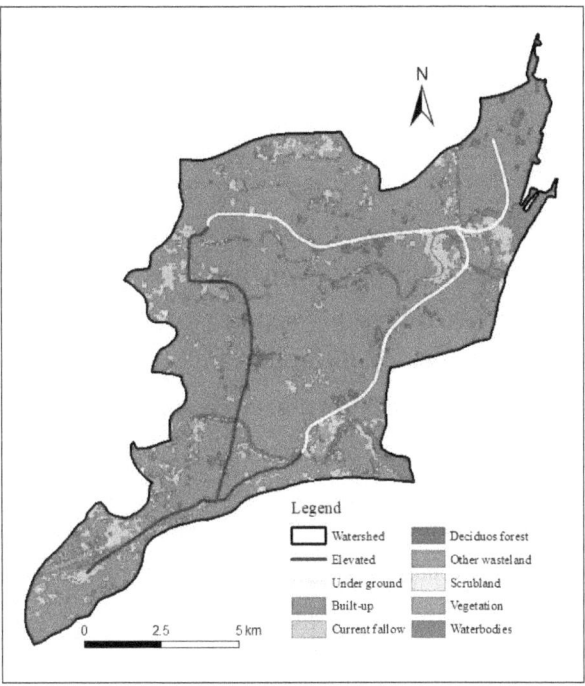

Figure 3.10 Land Use

The land use pattern of the underground stretch extends from East to West covering from Central to Anna Nagar comprising of Institutional and residential use and the tunneled stretch from North to South, that is from Washermenpet to Saidapet is mostly for Institutional, Commercial and Industrial use. Hence the demand of water for the stretch of Washermenpet to Saidapet will be more when compared with the stretch from Central to Anna Nagar.

3.14   SOCIAL AND CULTURAL DETAILS

Chennai is the fourth largest city in the country. Migrants in Chennai city account for nearly 40% of the population of the city. With recent developments in IT and ITES, the city employs about 1.8 million people, most of whom have come from various parts of the country, but mostly from Tamil Nadu. There are prospects for further expansion in the near future. The city is now a cultural melting pot with people from multiple ethnic-religious groups. The one reason for the industrial development in the city and the surroundings is the availability of infrastructures, besides, of course, industrial promotion through various means.

The infrastructures, particularly health, education, recreation, and housing, have made possible the incoming migrant flows in steady streams and the migrants have made the city their homes, for it provides for all their needs.

3.15   DATA COLLECTION

3.15.1   Rainfall Data

Rainfall is the major source of surface and underground water source. The pattern rainfall for various years was studied. The rainfall data for 22 years was collected and a plot was constructed. The resource generated due to the source of rainfall also makes an impact on the water levels. The ultimate recharge under the surface of the ground depends upon the highest precipitation occurring at the ground surface. Table 3.3 gives the Monthly Rainfall data from 1995-2017.The annual rainfall was seemed to be above 2000 mm in the years 1997,2005 and 2015 due to flood in those years.

Table 3.3 Monthly Rainfall data from 1995-2017

Year	Jan	Feb	March	April	May	June	July	Aug	Sept	Oct	Nov	Dec	Annual Rainfall (mm)
1995	171.2	0	0	0	280.1	19.1	101.7	238.7	190.8	249.1	274.3	0	1525
1996	0	0	0	15.8	15	697.5	65.4	0	0	0	330.3	442.5	1566.5
1997	8.3	0	0	50.2	0.6	61.4	50.9	100.4	193	276.1	831.9	442.7	2015.5
1998	3.5	0	3.8	17.3	19.9	43.5	90.4	113.6	66.2	187.1	353.1	180.7	1079.1
1999	9.6	0.2	0	16.9	12.1	117.8	100.1	189.1	175.1	311.2	168.6	50.4	1151.1
2000	0	212.8	0.6	19.8	21.2	87	126.8	53.6	162.5	162.8	190.2	42.3	1079.6
2001	1.4	0	0	72.4	19.6	20	269.5	47.9	125	411.4	370	288.4	1625.6
2002	39.1	5.6	0	0	17.2	34.2	80	99.9	137.9	375.1	580.3	29.9	1399.2
2003	0.1	0	6.2	0	0.5	22.9	100.6	142.1	155.1	149	85	77.4	738.9
2004	51.5	0	0.2	2.4	210.6	28.8	50.6	47.1	246.5	285.3	280.2	6.8	1210
2005	2	5.2	0	83	30.9	30.6	151.2	53.7	102.1	1077.8	608.4	421.5	2566.4
2006	3.5	0	9.5	7.1	17.2	36.4	72.1	140.8	143.9	633.9	238	20.8	1323.2
2007	0	6.6	0	0.2	0	94.2	243.9	170.9	167.7	274.9	95	256.3	1309.7
2008	50.2	10	137.9	26.9	0.3	126.4	28.6	147.2	120.9	372.9	556.6	17.7	1595.6
2009	21.1	0	2	0	14.7	22.5	36.8	87	86.8	71.8	562.6	275.2	1180.5
2010	5.2	0.4	0	0	204	136.7	155.2	254	120.1	195.7	274	287.5	1632.8
2011	10.8	88.9	0	18.5	12.6	130.2	65.8	368.9	286.2	260	457.2	134.8	1833.9
2012	16.3	0	0	0.2	0	24.7	79.89	121.9	214.1	422.6	47.7	125.5	1052.89
2013	0	14.3	11.7	3.6	11.9	34	146.6	195.1	240.1	157.2	193.6	85.9	1094
2014	27.3	34.6	4	12.2	39.4	71.1	121	138.7	161	373	409.9	152	1544.2
2015	9.47	0	0	61.05	9.37	36.73	165.475	137.35	78.775	143.325	1039.425	484.325	2165.295
2016	7.01	0.59	0.1	0.1	261.15	130.15	115.79	106.15	183.81	73.08	17.6	194.2	1089.73
2017	0	1	12	0.2	5.7	87.5	92.9	155.4	124.9	333	832.8	35.4	1680.8

Source:Institute of water studies,Tharamani,Chennai

3.15.2   Identification of Wells

To predict the changes in the water level below the surface of the ground the wells around the underground corridor had to be identified. Government sectors monitoring wells were selected to get the secondary water level dat. Nearly nine wells were selected around the two underground corridors. Secondary data were collected over a period of 21 years (1995-2015) from Central Ground Water Board and Institute for Water Studies, Chennai. The patterns of the wells with respect to the water level fluctuation were studied. Inspite of the secondary data primary data were observed for the past three years (2015-2017) by selecting 20 observation wells, 10 wells, 5 wells on each side for one corridor. The observation wells identified for both secondary and primary are listed in the Table 3.4 and the same is shown in the Figure 3.11.

Table 3.4 Observation Wells Selected around the Metro Rail Corridor

Secondary	Primary	
Tondiarpet	Bible Society	Chetpet
Vepery	Aziz Mulk	Tirumangalam(2)
Chepauk	Tirumangalam(2)	GAA
Thousand Lights	Teynampet	AnnaNagar(1)
Saidapet	DMS	AnnaNagar(2)
Guindy	Nandanum(1)	School Road
Aminjikarai	Nandanum(2)	Vanavil(1)
Tirumangalam	S.Market	Vanavil(2)
K.K.Nagar	Saidapet	AP Flates
Kilpauk	KMC	UPC

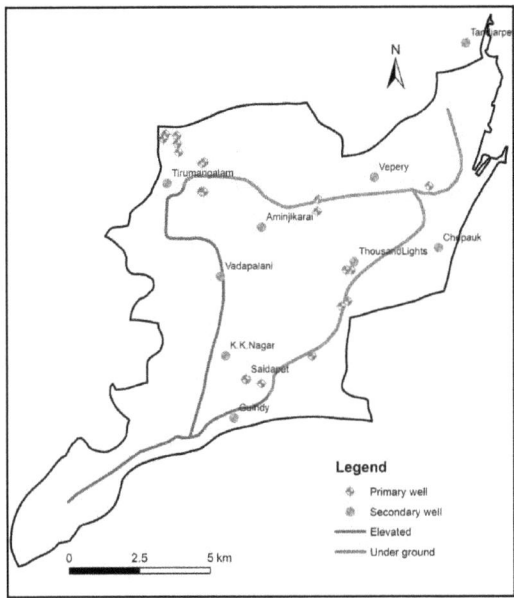

Figure 3.11 Observation Wells of the Under Ground Corridors

3.16   WATER LEVEL DATA

Water level data were collected from the government sectors for a period of 21 years from 1995 -2015 to observe the changes occurred before and during the construction of Metro rail corridor. The annual means of water level for the above mentioned period are given below.

Primary water level data were obtained with 20 wells being selected on either side of the underground stretch for a period from 2015-2017 to assess the changes that occurred after the construction of the corridors. Water levels were measured with the help of the thread. Water levels were calculated from the surface of the well to the tip of the surface level of the well. The raw data, both secondary and primary water levels were collected below ground level.

3.16.1   Secondary Wells Data

The Secondary wells data were collected for over a period of 21 years (1995-2015) from the Central Ground Water Board and the Institute for Water Studies sections for the observation wells around the metro rail corridors. On the shore side Thousand Lights, Chepauk, Tondiarpet, Guindy and Saidapet were identified as secondary wells for data collection. In the underground stretch from Central to Thirumangalam, Vepery, Aminijikarai, Thirumangalam and K.K. Nagar monitor wells were choosen for the data collection. Table 3.5 shows the average water level data collected for a period from 1995-2015 and it is also shown in the Figure 3.12.

3.16.2   Primary Wells Data

The Primary wells data were collected for over a period of 2 years (2015-2017) from the observation wells around the metro rail corridors. Five wells were located on either side of the corridor to assess the effect on both sides of the corridor. The observation located on the Central to Thirumangalam stretch are Chetpet, Tirumangalam(2), GAA, AnnaNagar(1), AnnaNagar(2), School Road, Vanavil(1),Vanavil(2), AP Flates and UPC. The observation located on the Central to Saidapet stretch are Bible Society, Aziz Mulk, Tirumangalam (2), Teynampet, DMS, Nandanam(1), Nandanam(2), S.Market, Saidapet, KMC. The Primary Observation wells with the water were levels monitored over the period from December 2015 to December 2017 . The water levels were monitored at a frequency of once in two months and are presented in Table 3.6.

Table 3.5 Water Level Data for the Secondary Wells from 1995-2015

Observation Well	1995	1996	1997	1998	1999	2000	2001	2002	2003	2004	2005
Tondiarpet	6.54	6.39	6.8	6.76	5.36	5.7	5.48	5.9	5.26	5.98	3.96
Vepery	9.17	7.48	6.09	5.87	7.49	5.54	4.05	6.6	7.23	5.04	4.96
Chepauk	8.39	8.6	8.47	8.36	8.33	8.19	8.12	8.56	7.68	8.41	8.48
ThousandLights	6.77	5.37	6.79	7.11	5.39	5.7	5.44	6.4	5.2	5.51	5.65
Saidapet	10.35	9.9	9.35	9.82	9.36	8.24	9.5	9.34	9.32	9.15	11.3
Aminjikarai	7.96	6.4	7.12	6.68	7.08	5.29	14.12	7.73	4.83	4.12	2.45
Tirumangalam	9.74	9.23	10.05	11.15	11.43	8.42	8.16	9.2	8.51	6.11	7.67
K.K.Nagar	13.75	12.84	13.2	13.3	13.33	12.46	10.46	13.07	11.8	8.97	10.67

Table 3.5 (Continued)

Observation Well	2006	2007	2008	2009	2010	2011	2012	2013	2014	2015
Tondiarpet	7.37	6.94	7.25	7.03	6.05	5.12	5.47	5.01	5.46	3.78
Vepery	5.92	5.75	4.51	4.19	3.71	3.63	4.15	3.18	3.46	3.41
Chepauk	8.37	7.33	7.32	6.67	6.45	7.08	7.09	6.44	6.26	5.80
ThousandLights	5.94	5.45	5.65	5.14	5.25	5.1	5.22	5.24	4.75	4.95
Saidapet	11.99	8.74	6.71	14.62	8.16	8.97	8.33	9.86	9.92	9.85
Aminjikarai	7.08	7.14	8.07	8.07	7.37	7.96	8.65	7.25	6.25	14.12
Tirumangalam	9.77	11.13	11.84	11.84	11.70	12.94	12.96	12.38	10.49	13.82
K.K.Nagar	14.22	14.29	13.36	13.36	12.54	13.70	14.77	13.85	13.54	14.75

Source: Institute for Water Studies & Central Ground Water Board, Chennai

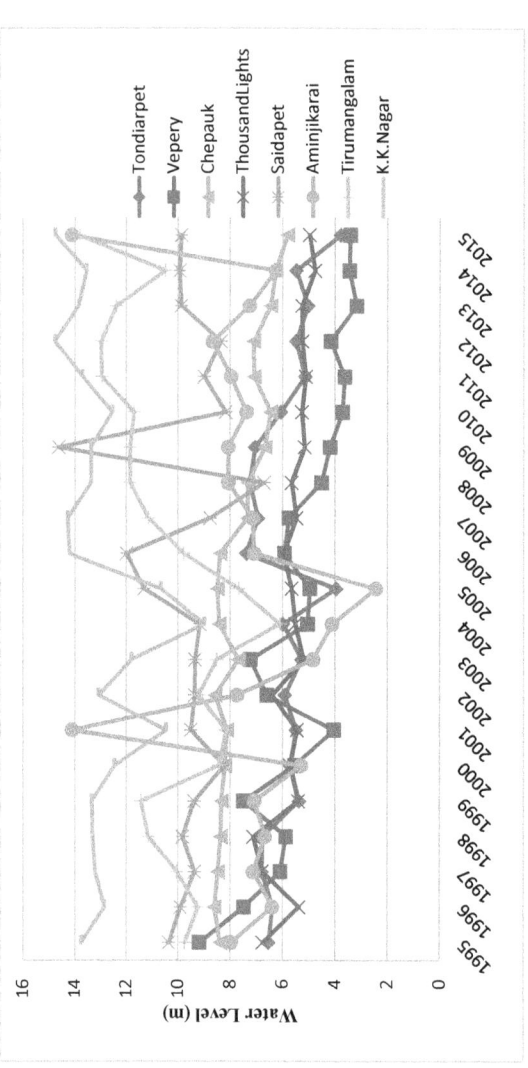

Figure 3.12 Water Levels in the Wells

The wells like Vepery, Thousand lights and Tondiarpet shows low water levels. The wells like K.K.Nagar, Thirumangalam and Saidapet shows the water levels in the higher range. The well Aminjikarai shows two peak values in the years 2001 and 2015.The well Chepauk does not shows any peak values and it seemed to be carried a decreased pattern from 1995-2015.

Table 3.6 Water Level Data for the Primary Wells from Dec -2015 to Dec-2017

Observation Well	Dec-2015	Feb-2016	Apr-2016	Jun-2016	Aug-2016	Oct-2016	Dec-2016	Feb-2017	Apr-2017	Jun-2017	Aug-2017	Oct-2017	Dec-17
Bible Society	3.47	3.52	4.46	5.26	8.54	6.5	5.54	5.32	5.18	4.99	4.6	4.6	4.54
Aziz Mulk 3rd st	6.43	6.49	5.61	7.79	11.59	9.29	7.29	7.03	6.96	6.89	6.7	6.57	6.34
GAA 9th street	5.19	6.69	8.64	10.85	11.49	10.73	9.54	9.22	8.99	8.8	8.63	8.46	8.29
Teynampet	5.67	5.09	5.19	7.52	11.79	10.69	9.34	9.01	8.96	8.79	8.59	8.29	7.99
DMS	2.84	2.44	2.44	5.34	8.74	7.14	6.2	5.87	5.74	5.41	5.24	4.99	4.64
Nandanam (1)	7.04	7.14	7.24	7.7	8.34	7.84	7.29	6.94	6.74	6.5	6.17	5.96	5.84
Nandanam (2)	4.64	5.14	5.34	7.79	8.34	7.86	6.79	6.52	6.44	6.2	5.97	5.91	5.74
Saidapet Jeenis road	6.51	6.56	7.63	8.39	1.49	9.65	8.99	8.72	8.53	8.22	7.74	7.06	6.59
Saidapet	3.94	4.14	4.89	5.69	8.44	6.34	5.74	5.33	5.2	4.99	4.5	4.42	4.24
KMC	6.85	6.74	7.29	10.95	1.59	10.69	10.19	9.89	9.74	9.59	9.25	9.03	8.54
Chetpet	6.01	5.99	6.29	9.52	1.79	10.25	8.59	8.04	7.93	7.86	7.54	7.22	7.02
Thirumangalam(1)	5.24	5.04	6.28	7.84	8.34	7.74	7.34	7.05	6.98	6.71	6.48	6.16	5.94
Thirumangalam(2)	5.14	4.14	5.49	6.99	8.54	7.42	6.94	6.55	6.47	6.41	6.29	5.97	5.44
Anna Nagar (1)	6.14	5.94	7.29	7.59	8.64	7.68	7.27	6.98	6.91	6.74	6.58	6.4	6.24
Anna Nagar (2)	5.34	5.14	6.5	7.16	8.74	7.57	6.93	6.54	6.48	6.36	6.3	6.17	6.14
Handicap School	5.67	5.5	6.78	7.54	8.34	7.7	7.34	7.08	6.89	6.58	6.17	5.84	5.64
Vanavil apart (1)	8.79	8.19	9.72	10.53	11.59	10.46	9.89	9.44	9.33	9.2	9.15	8.9	8.79
Vanavil apart (2)	8.89	8.69	8.09	10.64	11.59	11.26	10.31	10.14	10.03	9.73	9.42	9.26	8.89
Ap flats	6.44	6.24	6.94	7.56	8.3	7.5	7.34	7.18	7.1	7.03	6.89	6.82	6.74
United colony	5.04	4.96	5.41	6.41	8.34	6.5	5.94	5.46	7.08	7.04	6.57	6.38	6.14

3.17   WATER QUALITY DATA

To assess the water quality scenario before and after the construction of metro rail corridor the secondary data's were collected for the period of 21 years from Central Ground Water Board and Institute for Water Studies, Chennai. The primary data which were collected were divided into two phases of construction - before and after the construction of metro rail corridors and it were compared.

3.17.1   Secondary well concentrations

With regard to the Secondary wells listed above in Table 3.4, water quality data were collected from the respective government sectors. The data were collected for a period 21 years from 1995-2015 and the annual mean values are given in Table The important parameters like pH, Total Dissolved Solids (TDS), Total Hardness (TH), Chloride (Cl) and Fluoride (Fl) were considered for the analysis. The above mentioned parameter data were collected for all the observation wells taken and the sample data are presented in Table 3.7 for the wells Tondiarpet and Vepery.

3.17.2   Primary Well Concentrations

In order to evaluate the secondary data collection, the primary data were collected for the 20 wells selected around the corridors. The water samples were collected for almost 2 years and were analyzed as per the recommended procedures and the results were obtained. Table 3.8 furnished the concentrations levels measured in ppm of the primary wells selected on either side of the corridor.

**Table 3.7   The Concentrations of the secondary wells (a) Tondiarpet
(b)Vepery**

(a)   **Well Location –Tondiarpet**

Year	pH	TH	Ca	Cl	F	TDS
1996	7.95	165	52	227	-	578
1997	7.75	373	46	303	-	873
1998	8.4	530	84	305	-	957
1999	8.5	385	59	516	-	1434
2000	8.6	690	24	369	0.85	2010
2006	7.9	920	80	1647	0.62	-
2007	7.9	920	80	1647	0.62	-
2008	7.33	250	36	389	0.60	256
2009	6.99	205	56	92	0.92	262
2010	7.80	480	96	131	0.70	330
2011	7.94	370	68	177	0.78	329
2013	6.92	270	52	213	1.26	329

(b)   **Well Location –Vepery**

Year	pH	TH	Ca	Cl	F	TDS
1996	8.25	335	40	716	-	1638
1997	7.45	420	66	510	-	1555
1998	8.30	550	160	273	-	1234
1999	8.40	570	76	1018	-	2376
2000	8.10	450	32	468	0.75	1850
2001	7.90	1050	40	2978	0.71	6467
2002	8.10	240	64	138	0.10	444
2003	7.80	1150	80	262	1.09	1619
2004	7.30	850	60	1347	0.22	3804
2005	7.75	1100	120	1330	0.20	3484
2006	7.53	975	90	1338	0.21	3644
2007	7.64	1038	105	1334	0.21	3564
2008	7.30	580	44	688	1.52	720
2009	7.67	610	44	858	1.20	500
2010	7.80	1010	100	702	1.27	220

Source: Institute for Water Studies & Central Ground Water Board, Chennai

Table 3.8 The concentration of the primary wells

Well Location -Bible Society

Date	pH	TDS	EC	TH	Ca	Cl	Fl
Dec-15	9.4	1300	3.2	1290	530	1200	0.13
Feb-16	9.4	1300	3.2	1292	530	1200	0.13
Apr-16	9.4	1300	3.2	1292	525	1200	0.13
Jun-16	9.1	1299	3.0	1282	522.5	1198	0.11
Aug-16	9.1	1300	3.1	1290	530	1200	0.12
Oct-16	9.3	1305	3.1	1290	531	1205	0.13
Dec-16	9.4	1305	3.2	1290	530	1221	0.13
Feb-17	9.3	1310	3.3	1589	527	1225	0.15
Apr-17	9.5	1315	3.5	1590	527.5	1231	0.16
Jun-17	9.4	1315	3.6	1590	528	1225	0.15
Aug-17	9.4	1315	3.6	1590	528	1225	0.15
Oct-17	9.5	1320	3.7	1591	529	1228	0.16

Well Location -GAA 9th street

Date	pH	TDS	EC	TH	Ca	Cl	Fl
Dec-15	8.9	950	3.9	875	291.7	946	0.35
Feb-16	8.9	950	3.9	864	291.7	946	0.35
Apr-16	8.9	950	3.9	864	300.5	946	0.35
Jun-16	8.4	928	3.9	859	279	1023	0.25
Aug-16	8.9	950	3.9	861	281.7	946	0.35
Oct-16	8.9	950	3.9	861	291.7	946	0.35
Dec-16	8.9	950	3.9	867	291.7	946	0.35
Feb-17	8.3	932	4.0	868	292	1037	0.26
Apr-17	8.4	933	3.9	868	292.9	1038	0.27
Jun-17	8.1	934	4.0	869	293	1038	0.25
Aug-17	8.1	934	4.0	869	293	1038	0.25
Oct-17	8.2	935	3.9	870.5	292	1040	0.24

Table 3.8 (Continued)

Well Location -Aziz Mulk 3rd st

Date	pH	TDS	EC	TH	Ca	Cl	Fl
Dec-15	8.3	1399	0.9	865	288.3	490	0.38
Feb-16	8.3	1400	0.9	967	322.3	490	0.38
Apr-16	8.6	1400	0.9	967	322.3	490	0.38
Jun-16	8.4	1369	0.7	952	317.3	522	0.26
Aug-16	8.2	1373	0.7	865	318	490	0.38
Oct-16	8.3	1380	0.9	865	320	490	0.38
Dec-16	8.3	1381	0.8	865	323.9	490	0.38
Feb-17	8.5	1381	1.1	960	324	528	0.30
Apr-17	8.7	1384	1.1	965	325	529	0.38
Jun-17	8.7	1385	1.3	966	325.9	530	0.38
Aug-17	8.7	1385	1.3	966	325.9	530	0.38
Oct-17	8.8	1386	1.4	967	325.9	532	0.38

Well Location -DMS

Date	pH	TDS	EC	TH	Ca	Cl	Fl
Dec-15	7.6	1000	3.7	478	159.3	479	0.36
Feb-16	7.6	1000	3.7	478	159.3	479	0.36
Apr-16	7.6	1000	3.7	478	159.3	479	0.36
Jun-16	7.1	925	3.7	460	153.3	456	0.32
Aug-16	7.6	1000	3.7	478	159.3	479	0.36
Oct-16	7.6	1000	3.7	478	159.3	479	0.36
Dec-16	7.6	1000	3.7	478	159.3	479	0.36
Feb-17	7.3	928	3.9	469	153	489	0.35
Apr-17	7.6	929	3.9	470	160	490	0.36
Jun-17	7.6	929	3.8	470	159	490	0.36
Aug-17	8	929	4	470	159	495	0.34
Oct-17	8	929	4	470	159	495	0.34

Well Location -Teynampet

Date	pH	TDS	EC	TH	Ca	Cl	Fl
Dec-15	8.1	350	3.2	1085	361.7	1150	0.56
Feb-16	8.1	350	3.2	1085	361.7	1150	0.56
Apr-16	8.1	350	3.2	1085	361.7	1150	0.56
Jun-16	7.9	342	3.2	1082	360.7	1234	0.38
Aug-16	8.1	350	3.2	1085	361.7	1150	0.56
Oct-16	8.1	350	3.2	1085	361.7	1150	0.56
Dec-16	8.1	350	3.2	1085	361.7	1150	0.56
Feb-17	8.1	345	3.8	1088	361	1239	0.42
Apr-17	8.2	351	3.8	1089	365	1239	0.42
Jun-17	8.2	350	3.9	1089	361	1240	0.43
Aug-17	8.2	350	3.9	1089	361	1240	0.43
Oct-17	8.1	351	3.8	1087	362	1241	0.42

Well Location -Nandanam-2

Date	pH	TDS	EC	TH	Ca	Cl	Fl
Dec-15	7.1	1800	1.2	350	116.7	258	0.48
Feb-16	7.1	1800	1.2	348	116	1026	0.48
Apr-16	7.1	1800	1.2	348	116	1026	0.48
Jun-16	7.1	1723	1.2	342	114	1000	0.39
Aug-16	7.1	1800	1.2	350	116.7	258	0.48
Oct-16	7.1	1800	1.2	350	116.7	258	0.48
Dec-16	7.1	1800	1.2	350	116.7	258	0.48
Feb-17	7.5	1733	1.5	348	114	1054	0.41
Apr-17	7.5	1734	1.5	350	115	1055	0.42
Jun-17	7.5	1734	1.6	350	115.9	1054	0.44
Aug-17	7.5	1734	1.6	350	115.9	1054	0.44
Oct-17	7.6	1735	1.7	355	116	1055	0.45

Table 3.8 (Continued)

Well Location -Nandanam-1

Date	pH	TDS	EC	TH	Ca	Cl	Fl
Dec-15	7.1	1800	1.2	348	115.2	1026	0.42
Feb-16	7.1	1800	1.2	350	116.7	258	0.42
Apr-16	7.1	1800	1.2	350	116.7	258	0.42
Jun-16	7.1	1721	1.2	349	116.3	234	0.42
Aug-16	7.1	1800	1.2	348	115.2	1026	0.42
Oct-16	7.1	1800	1.2	348	115.2	1026	0.42
Dec-16	7.1	1800	1.2	348	115.2	1026	0.42
Feb-17	7.1	1733	1.3	351	116	1028	0.44
Apr-17	7.2	1732	1.3	351	116	1029	0.44
Jun-17	7.2	1733	1.3	352	117	1030	0.45
Aug-17	7.2	1733	1.3	352	117	1030	0.45
Oct-17	7.2	1732	1.3	352	118	1030	0.45

Well Location -Saidapet Jeenis Road

Date	pH	TDS	EC	TH	Ca	Cl	Fl
Dec-15	8.3	950	3.9	1089	363	1388	0.32
Feb-16	8.3	950	3.9	1089	363	1388	0.32
Apr-16	8.3	950	3.9	1089	363	1388	0.32
Jun-16	7.9	943	3.9	1072	357.3	1250	0.28
Aug-16	8.3	950	3.9	1089	363	1388	0.32
Oct-16	8.3	950	3.9	1089	363	1388	0.32
Dec-16	8.3	950	3.9	1089	363	1388	0.32
Feb-17	8.1	952	3.9	1079	357	1243	0.3
Apr-17	8.3	955	4	1081	358	1244	0.31
Jun-17	8.3	953	3.9	1081	359	1245	0.32
Aug-17	8.3	953	3.9	1081	359	1245	0.32
Oct-17	8.4	954	4	1082	360	1245	0.32

Well Location -Saidapet

Date	pH	TDS	EC	TH	Ca	Cl	Fl
Dec-15	8.5	960	3.9	1068	356	1250	0.38
Feb-16	8.5	960	3.9	1068	356	1250	0.38
Apr-16	8.5	960	3.9	1068	356	1250	0.38
Jun-16	8.1	959	3.9	1054	351.3	1200	0.32
Aug-16	8.5	960	3.9	1068	356	1250	0.38
Oct-16	8.5	960	3.9	1068	356	1250	0.38
Dec-16	8.5	960	3.9	1068	356	1250	0.38
Feb-17	8.3	962	3.5	1056	351	1223	0.42
Apr-17	8.3	960	3.5	1055	351	1223	0.42
Jun-17	8.3	961	3.5	1056	356	1230	0.42
Aug-17	8.3	961	3.5	1056	356	1230	0.42
Oct-17	8.4	962	3.6	1058	357	1231	0.43

Well Location -Chetpet

Date	pH	TDS	EC	TH	Ca	Cl	Fl
Dec-15	8.1	1000	3.2	1043	347.7	250	0.86
Feb-16	8.1	1000	3.2	1043	347.7	250	0.86
Apr-16	8.1	1000	3.2	1043	347.7	250	0.86
Jun-16	7.9	981	3.2	1012	337.3	223	0.71
Aug-16	8.1	1000	3.2	1043	347.7	250	0.86
Oct-16	8.1	1000	3.2	1043	347.7	250	0.86
Dec-16	8.1	1000	3.2	1043	347.7	250	0.86
Feb-17	7.8	985	3.6	1019	337	321	0.74
Apr-17	7.9	985	3.3	1019	337	321	0.74
Jun-17	7.9	985	3.3	1019	337	321	0.74
Aug-17	7.9	985	3.3	1019	337	321	0.74
Oct-17	8	986	3.4	1020	338	322	0.75

Table 3.8 (Continued)

Well Location - KMC

Date	pH	TDS	EC	TH	Ca	Cl	Fl
Dec-15	8.4	470	3.7	1189	396.3	404	0.53
Feb-16	8.4	470	3.7	1189	396.3	404	0.53
Apr-16	8.4	470	3.7	1189	396.3	404	0.53
Jun-16	8.1	466	3.7	1174	391.3	389	0.48
Aug-16	8.4	470	3.7	1189	396.3	404	0.53
Oct-16	8.4	470	3.7	1189	396.3	404	0.53
Dec-16	8.4	470	3.7	1189	396.3	404	0.53
Feb-17	8.2	469	3.7	1178	371	412	0.52
Apr-17	8.4	470	3.7	1178	372	413	0.51
Jun-17	8.4	470	3.8	1179	372	412	0.51
Aug-17	8.4	470	3.8	1179	372	412	0.51
Oct-17	8.5	471	3.9	1180	373	415	0.52

Well Location - Thirumangalam-1

Date	pH	TDS	EC	TH	Ca	Cl	Fl
Dec-15	7.9	610	3.8	578	192.7	638	0.46
Feb-16	7.9	610	3.8	578	192.7	638	0.46
Apr-16	7.9	610	3.8	578	192.7	638	0.46
Jun-16	7.1	593	3.8	565	218.7	612	0.35
Aug-16	7.9	610	3.8	578	192.7	638	0.46
Oct-16	7.9	610	3.8	578	192.7	638	0.46
Dec-16	7.9	610	3.8	578	192.7	638	0.46
Feb-17	7.0	621	3.6	568	219	624	0.37
Apr-17	7.1	623	3.6	568	219	624	0.37
Jun-17	7.1	627	3.6	568	219	624	0.37
Aug-17	7.1	627	3.6	568	219	624	0.37
Oct-17	7.3	629	3.8	569	220	625	0.38

Well Location - Thirumangalam-2

Date	pH	TDS	EC	TH	Ca	Cl	Fl
Dec-15	7.9	610	3.8	578	192.7	1789	0.36
Feb-16	7.9	610	3.8	578	192.7	1789	0.36
Apr-16	7.9	610	3.8	578	192.7	1789	0.36
Jun-16	7.6	592	3.8	569	189.7	1588	0.27
Aug-16	7.9	610	3.8	578	192.7	1789	0.36
Oct-16	7.9	610	3.8	578	192.7	1789	0.36
Dec-16	7.9	610	3.8	578	192.7	1789	0.36
Feb-17	7.9	578	3.8	571	190	1532	0.3
Apr-17	8.1	585	3.8	571	190	1532	0.3
Jun-17	8.1	590	3.8	571	190	1532	0.3
Aug-17	8.1	590	3.8	571	190	1532	0.3
Oct-17	8.2	591	3.9	572	191	1533	0.4

Well Location - Anna Nagar -2

Date	pH	TDS	EC	TH	Ca	Cl	Fl
Dec-15	7.1	895	1.0	644	214.7	970	0.86
Feb-16	7.1	895	1.0	644	214.7	970	0.86
Apr-16	7.1	895	1.0	644	214.7	970	0.86
Jun-16	7.1	892	1.0	639	213	956	0.73
Aug-16	7.1	895	1.0	644	214.1	960	0.86
Oct-16	7.1	895	1.0	646	214.9	962	0.86
Dec-16	7.1	895	1.0	650	214.7	965	0.86
Feb-17	7.3	892	1.4	649	215	942	0.70
Apr-17	7.3	893	1.3	650	214	967	0.71
Jun-17	7.4	895	1.4	651	215	969	0.71
Aug-17	7.4	895	1.4	651	215	969	0.71
Oct-17	7.4	895	1.4	651	215	969	0.71

Table 3.8 (Continued)

Well Location - Anna Nagar -1

Date	pH	TDS	EC	TH	Ca	Cl	Fl
Dec-15	7.1	890	1.0	644	214.7	876	0.57
Feb-16	7.1	890	1.0	644	214.7	876	0.57
Apr-16	7.1	890	1.0	644	214.7	876	0.57
Jun-16	7.1	879	1.0	633	211	834	0.45
Aug-16	7.1	890	1.0	644	214.7	876	0.57
Oct-16	7.1	890	1.0	644	214.7	876	0.57
Dec-16	7.1	890	1.0	644	214.7	876	0.57
Feb-17	7.1	879	1.1	683	217	826	0.46
Apr-17	7.2	889	1.1	683	217	826	0.46
Jun-17	7.3	880	1.1	683	217	826	0.46
Aug-17	7.3	880	1.1	683	217	826	0.46
Oct-17	7.4	881	1.2	684	218	827	0.48

Well Location - Handicap School

Date	pH	TDS	EC	TH	Ca	Cl	Fl
Dec-15	7.1	850	1.6	556	185.3	1074	0.34
Feb-16	7.1	850	1.6	556	185.3	1074	0.34
Apr-16	7.1	850	1.6	556	185.3	1074	0.34
Jun-16	7.1	848	1.2	551	183.7	1043	0.21
Aug-16	7.1	850	1.5	553	184	1052	0.34
Oct-16	7.1	850	1.6	556	185.3	1060	0.34
Dec-16	7.1	850	1.6	556	185.9	1064	0.34
Feb-17	7.3	853	1.8	562	186	1069	0.29
Apr-17	7.5	855	1.8	570	186.1	1070	0.30
Jun-17	7.5	855	2.0	572	188	1073	0.32
Aug-17	7.5	855	2.0	572	188	1073	0.32
Oct-17	7.6	856	2.1	573	189	1074	0.31

Well Location - Vanavil apart -1

Date	pH	TDS	EC	TH	Ca	Cl	Fl
Dec-15	7.6	780	2.4	639	213	672	0.57
Feb-16	7.6	780	2.4	639	213	672	0.57
Apr-16	7.6	780	2.4	639	213	672	0.57
Jun-16	7.1	779	2.4	633	211	652	0.49
Aug-16	7.6	780	2.4	639	213	672	0.57
Oct-16	7.6	780	2.4	639	213	672	0.57
Dec-16	7.6	780	2.4	639	213	672	0.57
Feb-17	7.6	785	2.3	639	215	667	0.44
Apr-17	7.7	786	2.4	641	216	670	0.46
Jun-17	7.7	786	2.5	645	220	672	0.47
Aug-17	7.7	786	2.5	645	220	672	0.47
Oct-17	7.8	787	2.6	646	222	674	0.43

Well Location - Ap Flats

Date	pH	TDS	EC	TH	Ca	Cl	Fl
Dec-15	7.1	820	3.6	786	262	1093	0.84
Feb-16	7.1	820	3.6	786	262	1093	0.84
Apr-16	7.1	820	3.6	786	262	1093	0.84
Jun-16	7.1	816	3.6	781	260.3	1079	0.79
Aug-16	7.1	820	3.6	786	262	1093	0.84
Oct-16	7.1	820	3.6	786	262	1093	0.84
Dec-16	7.1	820	3.6	786	262	1093	0.84
Feb-17	7.1	821	3.8	719	260	1080	0.80
Apr-17	7.2	823	3.7	788	265	1090	0.85
Jun-17	7.2	825	3.8	788	265	1095	0.86
Aug-17	7.2	825	3.8	788	265	1095	0.86
Oct-17	7.2	825	3.8	788	265	1095	0.86

Table 3.8 (Continued)

Well Location - United Colony

Date	pH	TDS	EC	TH	Ca	Cl	Fl
Dec-15	7.9	689	1.2	690	230	730	0.32
Feb-16	7.9	689	1.2	690	230	730	0.32
Apr-16	7.9	689	1.2	690	230	730	0.32
Jun-16	7.6	687	1.2	688	229.3	721	0.26
Aug-16	7.9	689	1.2	690	230	730	0.32
Oct-16	7.9	689	1.2	690	230	730	0.32
Dec-16	7.9	689	1.2	690	230	730	0.32
Feb-17	7.6	690	1.4	688	229	732	0.27
Apr-17	7.6	695	1.4	690	233.1	732	0.33
Jun-17	7.8	699	1.4	690	233	732	0.32
Aug-17	7.8	699	1.4	690	233	732	0.32
Oct-17	7.8	699	1.4	690	233	732	0.32

Well Location - Vanavil apart -2

Date	pH	TDS	EC	TH	Ca	Cl	Fl
Dec-15	7.6	780	3.2	635	211.7	794	0.38
Feb-16	7.6	780	3.2	635	211.7	794	0.38
Apr-16	7.6	780	3.2	635	211.7	794	0.38
Jun-16	7.1	769	3.2	632	210.7	782	0.26
Aug-16	7.6	780	3.2	635	211.7	794	0.38
Oct-16	7.6	780	3.2	635	211.7	794	0.38
Dec-16	7.6	780	3.2	635	211.7	794	0.38
Feb-17	7.5	772	3.2	639	211	790	0.29
Apr-17	7.6	773	3.2	640	212.5	791	0.37
Jun-17	7.6	780	3.3	643	212	797	0.38
Aug-17	7.6	780	3.3	643	212	797	0.38
Oct-17	7.6	780	3.3	643	212	797	0.38

The data which are collected above from the various laboratory test are used to analyze the effect of tunneling on either side of the Corridors I and II. These data are used as inputs in the water quality modeling to predict the changes in the concentration in the different stages of construction of tunneling Scenario.

3.18   SUMMARY

In this chapter, the complete knowledge about the study that was acquired with the thorough study of its climate, Rainfall pattern, Soil, Geology, Geomorphology, Hydrogeology, Land Use and slope has been discussed. The water level and water quality data were collected over the period of 21 years from 1995-2015 from the Central Ground Water Board and Institute for Water Studies. The primary well locations were identified opposite the underground sections and the data were collected for the period 2 years from 2015-2017. All the annual mean values of both water level and water quality have been presented above in the appropriate tabular columns.

CHAPTER 4

METHODOLOGY

4.1   INTRODUCTION

Construction of tunneling actually behaves as a barrier for the movement of groundwater. This causes changes in the ground water level and changes in the ground water quality too. Water level and water quality parameters data that were collected from the sources were used for the analysis and to predict the changes in the water level and the water quality. The predicted results are interpreted with the people survey is established. The detailed methodology is presented below in the flowchart.

4.2   METHODOLOGY FLOW CHART

The methodology starts with the collection of water levels from the observation wells located around the metro rail corridor underground section. The wells were identified and the water levels for the observation were collected for over a period of 22 years (1995-2017) from the PWD sections like Central Ground Water Board, Adyar and Institute for Water Studies, Taramani. From the same Centers, the water quality parameters data were collected for the same duration mentioned above to analyze the water quality standards below the ground surface around the metro rail corridors. The data are separated into periods before the tunnel construction and after the construction. Spatial analysis of both water level and water quality parameters was done using the GIS Software to assess the scenario in a graphical representation.

In the close proximity of the underground metro rail corridor stretch, five observation wells are located on either side of the corridor stretch to analyze the exact ground water table depth and the concentrations around the corridor. The water levels and the sampling procedure were carried out for over a period of two years 2015-2017. The spatial analysis for the primary wells was also done using GIS adopting inverse distance weighing interpolation technique. The comparison of the concentrations in each well was also done. The vector analysis was done by buffering the metro rail for the proximity of 2 km on either side of the corridor. Water Quality Index methodology was adopted to assess the status of the water quality standards around the metro rail corridors.

The water level data and the water quality data collected were used for the water flow modeling and the contaminant transport modeling. The various inputs required for the modeling such as rainfall, recharge, soil types, lithological data, permeability data and the geological data were collected for the flow model. Head Observation wells and the pumping wells are located in the base map were uploaded in the flow model. Model calibration, model validation and the model prediction were done for the progress of tunneling in the study area to assess the changes in the direction of flow due to the barrier created in the form of tunneling. The same was done for the selective parameters of the water quality and the changes were observed for the prediction of the years up to 2020. The predicted results of both water level and concentration were compared with the public hearing. The methodology flow chart is shown in Figure 4.1.

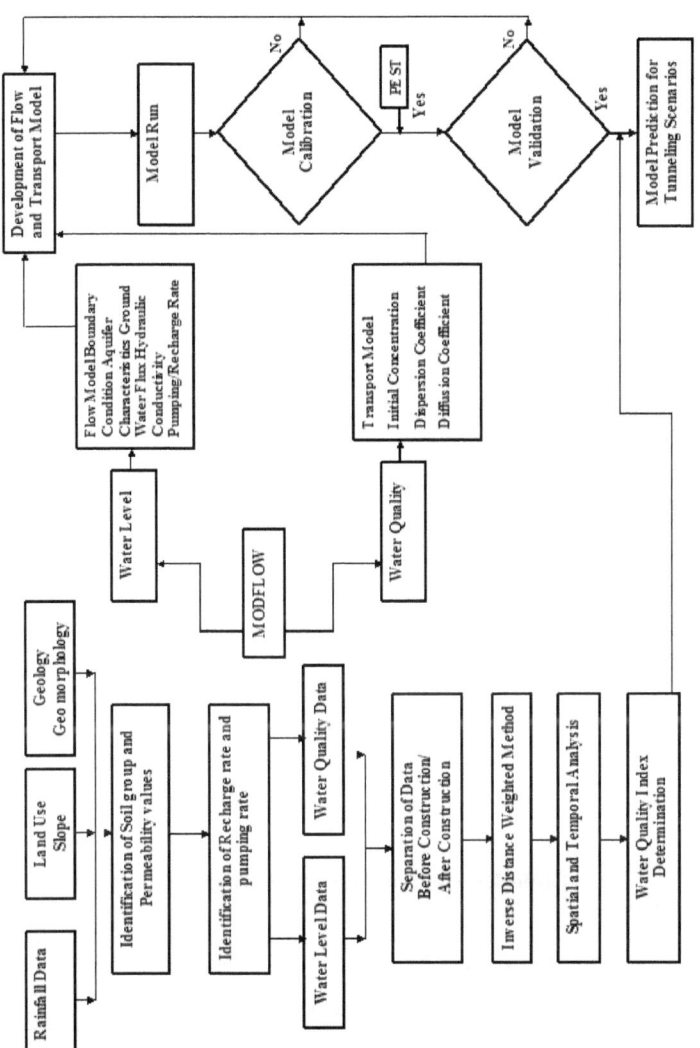

Figure 4.1 Methodology Flow Chart

4.3   WATER QUALITY

The construction of the metro rail corridor below the ground surface actually breaks out the entire heterogeneous character of the water quality profile below the surface of the earth. The basic parameters which determine the water quality have been taken and also some of the parameters which are likely to get changed due to the construction. Five parameters have been identified including the basic parameters and the parameters which are likely to affect purely due to the construction methodology and the procedure adopted for Chloride, Fluoride, pH, Total Dissolved Solids and Total Hardness.

4.3.1   Water Quality Sampling Procedure

The well water was collected either through pumping in wells with or without pumps and samples were collected from the bottom of the wells in clean polyethylene bottles, transported by placing in cooled ice boxes and refrigerated until analysis at 4°C. The samples were collected separately for physico-chemical analysis. The physical parameters considered for the study included pH, Total Dissolved Solids (TDS), and Total Hardness (TH). The chemical parameters included anions such as Fluoride (Fl) and Chloride (Cl). The methods used for the analysis are given in Table 4.1.

Table 4.1 Methodology of Analysis

S.No.	Quality Parameter	Method
1	pH	pH meter
2	Total Hardness	EDTA Titration
3	Chloride	Silver Nitrate Titration
4	Fluoride	Ion Selective Electrode Method
5	Total Dissolved Solids	Electrical Conductivity

4.4 WATER QUALITY INDEX

Water quality index determination is a widely accepted methodology to rate the water quality. WQI is defined as a rating reflecting the composite influence of different water quality parameters. Water quality index has been found to be the most effective to the citizens concerned and policy makers as it serves to communicate information on the quality of water. Water Quality Index (WQI) represents the gradation in water quality. It is defined as a rating technique as it reflects the composite influence of different water quality parameters. WQI is calculated to check the suitability of groundwater from the point of view of human consumption.

4.4.1 Water Quality Index Determination Procedure

The selected parameters are assigned with a weight relative to its importance in the drinking water quality. Parameters such as pH, EC, TDS, F, and so$_4$ are given much importance and assigned with the value 4. Total Hardness, Total Alkalinity and chloride are assigned with 3. Calcium and Magnesium carry 2. The relative weight (W_i) is computed using a weighted arithmetic index method given below (Brown $et\ al.$ 1972; Horton 1965; Tiwari & Manzoor 1988) in the following steps.

$$W_i \ = \ w_i \ / \ \Sigma \ w_i \tag{4.1}$$

where, W_i is the relative weight, w_i is the weight of each parameter and n is the number of parameters.

$$\text{Quality rating Scale} \ (Q_i) = \frac{\text{Concentration of each water sample } (mg/l)}{\text{Respective standard as per guidelines BIS 10500}}$$

$$\tag{4.2}$$

The Sub Index (SI$_i$) is calculated using the formula SI$_i$ = W$_i$ × Q$_i$ for each parameter. The overall Water Quality Index (WQI) is calculated by adding together each sub index value of each groundwater sample as follows:

WQI =Σ SI$_i$. Water quality classification based on WQI value is shown in Table 4.2.

Table 4.2 Water Quality Classification based on WQI Value

Class	WQI Value	Water Quality Status
I	<50	Excellent
II	50-100	Good Water
III	100-200	Poor Water
IV	200-300	Very poor Water
V	>300	Water unsuitable for Drinking

4.5   GIS ANALYSIS METHODOLOGY

A "Geographic Information System" (GIS) is a computer-based tool that allows one to create, manipulate, analyze, store and display information based on its location. GIS makes it possible to integrate different kinds of geographic information, such as digital maps, aerial photographs, satellite images and Global Positioning System data (GPS), along with associated tabular database information. GIS stands for Geographical Information System. It is defined as an integrated tool, capable of mapping, analyzing, manipulating and storing geographical data in order to provide solutions to real world problems and help in planning for the future. GIS deals with what and where components of occurrences. GIS allows one to perform statistical analysis or spatial queries, to explore 'what-if' scenarios, and to create predictive models.

4.5.1   Spatial Analysis Interpolation

The general formulation of the spatial interpolation problem can be defined as follows: Given the N values of a studied phenomenon z_j, $j = 1, \dots, N$ measured at discrete points $r_j = (x_j^{[1]}, x_j^{[2]}, \dots, x_j^{[d]})$, $j = 1, \dots, N$ within a certain region of a d-dimensional space, find a d-variate function $F(r)$ which passes through the given points, that means, fulfils the condition

$$F(r_j) = z_j, \qquad j = 1, \dots, N \tag{4.3}$$

Because, there exist an infinite number of functions which fulfill this requirement, additional conditions have to be imposed, defining the character of various interpolation techniques. Typical examples are conditions based on geostatistical concepts (Kriging), locality (nearest neighbor and finite element methods), smoothness and tension (splines), or ad hoc functional forms (polynomials, multi-quadrics). Choice of the additional condition depends on the character of the modeled phenomenon and the type of application.

4.5.1.1   Inverse distance weighing method

This is one of the simplest and most readily available methods. It is based on the an assumption that the value at an unsampled point can be approximated as a weighted average of values at points within a certain cut-off distance, or from a given number m of the closest points (typically 10 to 30). Weights are usually inversely proportional to a power of distance (Burrough 1986; Watson 1992) which, at an unsampled location r, leads to an estimator

$$F(r) = \sum_{r-1}^{m} w_i = (r_i) = \frac{\sum_{i=1}^{m} (r_t)/|r - r_i|^p}{\sum_{j=1}^{m} 1/|r - r_i|^p}$$

(4.4)

where p is a parameter (typically p=2; for more details on the influence of this parameter see Watson 1992). While this basic method is easy to implement and is available in almost any GIS, it has some wellknown shortcomings that limit its practical applications (Burrough 1986; Franke & Nielson 1991; Watson 1992). The method often does not reproduce the local shape implied by data and produces local extreme at the data points (Plate 26 (c)). A number of enhancements have been suggested, leading to a class of multivariate blended IDW surfaces and volumes (Franke & Nielson 1991; Tobler & Kennedy 1985; Watson 1992). However, most of these modifications are not implemented within GIS.

4.5.1.2   Spatial query-vector analysis

A spatial query is a special type of database query supported by geo databases. The queries differ from SQL queries in several important ways. Two of the most important are that they allow for the use of geometry data types such as points, lines and polygons and that these queries consider the spatial relationship between these geometries. Queries include both comparison (=, >, <, >=, <=, <>) and Boolean (AND, OR and NOT) operators.

4.6   VISUAL MOD FLOW

Groundwater models are computer models that provide a simplified representation of the processes that occur in the natural groundwater environment. Models are tools used by hydro geologists to simulate and predict aquifer conditions. A groundwater model can help you to make predictions about the behavior of the groundwater flow system: A groundwater model can incorporate all of these complexities, and assess different options and future conditions. When you are developing a groundwater model, it is necessary to translate the physical world into the modeling program. Geology becomes the hydro geologic parameters such as conductivity and storativity.

Hydrologic boundaries that impact the groundwater flow system are known as boundary conditions in a model, and include areas of recharge, rivers, lakes, wells, etc. In the physical world, one has field observations such as groundwater levels, fluxes, or contaminant concentrations and these are used to calibrate the model, making the model most closely match what is observed in the real world. "MODFLOW is developed by the U.S. Geological Survey (USGS); it is a three-dimensional (3D) finite-difference groundwater model. MODFLOW is considered an international standard for simulating and predicting groundwater conditions and groundwater/surface-water interactions". MODFLOW has been used for more than 30 years, and is widely accepted for its ease of use and flexibility in working with other programs.

4.6.1 Governing Equation of Ground Water Flow and Contaminant Transport

4.6.1.1 Ground water flow equation

Mathematical models provide a quantitative framework for analyzing data from monitoring and assessing quantitatively responses of the groundwater systems subjected to external stresses. Over the last four decades there has been a continuous improvement in the development of numerical groundwater models (Mohan 2001). Groundwater modeling is a powerful management tool which can serve multiple purposes such as providing a framework for organizing hydrologic data, quantifying the properties and behavior of the system and allowing quantitative 79 predictions of the responses of those systems to externally applied stresses (Senthilkumar & Elango 2004).

A computer program or code solves a set of algebraic equations generated by approximating the partial differential equations that form the mathematical models. The hydraulic head is obtained from the solution of three dimensional groundwater flow equation through MODFLOW soft ware (McDonald & Harbaugh 1988). Anisotropic and heterogeneous three-dimensional flow of groundwater, assumed to have constant density, may be described by the partial-differential equation:

$$\frac{d}{dx}\left[K_{xx}\frac{dh}{dx}\right] + \frac{d}{dy}\left[K_{yy}\frac{dh}{dy}\right] + \frac{d}{dz}\left[K_{zz}\frac{dh}{dz}\right] - W = S_s\frac{dh}{dt}$$

(4.5)

where, K_{xx}, K_{yy}, K_{zz} are components of the hydraulic conductivity tensor, h is potentiometric head, W is source or sink term, Ss is specific storage, and t is time. The finite-difference computer code Visual MODFLOW (McDonald & Harbaugh 1988) numerically approximates this equation, and was used to simulate the groundwater flow in the study area.

4.6.1.2   General mass balance equation

The equation governing the movement of the dissolved constituents in groundwater due to advection and dispersion can be developed, by utilizing a conservation of mass approach and employing Fick's law of dispersion. The mass balance of pollutant transport can be stated as,

$$\sum I + \sum P - \sum O - \sum L = \sum A$$

(4.6)

where I - input, P - production, O - output, L - loss, A – Accumulation

4.7 PROCEDURAL STEPS FOR VISUAL MODFLOW

4.7.1 Model Geometry

The base map or any image of the study area with proper resolution has to be uploaded for proper clarity. Minimum of 10000 m x 10000 m is given as input for the visual mod flow. It was uploaded with the observation wells and pumping wells.

4.7.2 Hydro Geological System

The lithological data from the PWD data and the data collected from the private house hold and construction agencies are considered for layer classification. Based on the geology, lithological stratifications and the mode of occurrence of groundwater conditions in different geological formations, the extent of the aquifer and the hydro geological conditions are classified for the study. The elevation for the control points with respect to the MSL has been taken to model the aquifer.

4.7.3 Aquifer Characteristics

The permeability values are well studied for the types of soil available in the study area and also from the literature surveys. It is a very important factor in the flow model to reduce the uncertainties in the velocity estimate. It also influences the contaminant fate and transport .Hence the permeability plays prominent role in the both flow and the contaminant transport modeling.

4.7.4 Boundary Condition

Initial condition specifies the value of the dependent variable, head, at all points within the model domain at some initial time, usually taken as $t = 0$.

The initial head values are also used to calculate the drawdown values, as measured by the difference between the starting head and the calculated head. The initial conditions are considered from the water level, and the water quality data obtained from the observation wells, from the initial year of modeling.

Boundary conditions are mathematical statements, which specify the dependent variable at the boundaries of the problem domain. A boundary condition that specifies the value of a state variable, i.e., the head along a boundary segment, is the Dirichlet boundary condition (Bear & Cheng 2009) considered for the study. The boundary conditions are specified based on the hydrogeology conditions of the study area.

4.7.5   Modeling Process

4.7.5.1   Model calibration

Model calibration is an important phase and a difficult task in the modeling process. Model calibrations minimize the deviation between the observed field data and the simulated data. It take several iterative stages to achieve the goal.

Hydraulic conductivity, storage coefficient, and input and output stresses are adjusted in the iterative process to achieve the model calibration. The accuracy of the computed groundwater head and the concentration was judged by a mean error, mean absolute error and root mean square error.

4.7.5.2   Model validation

The selected values of the hydrological parameters were used in the Model calibration and further refined in the Model validation process. If the model is successfully producing the measured changes, then the model is ready for the prediction process.

4.7.5.3   Model prediction

Model prediction is done for a various different scenarios like changes in the pumping rate, recharge rate etc. If the model calibration and the model validation are done properly, then the prediction will give a higher level of accuracy. The various management decisions can be well made with the prediction process.

4.7.6   Solute Transport Equation

MT3DMS is a transport model for simulating advection, dispersion, and chemical reactions of contaminants in groundwater flow systems. It solves the transport equation after the flow solution has been obtained from the groundwater flow model (MODFLOW). The general adjective-dispersive equation describing the fate and transport of the contaminant of species k in three-dimensional transient groundwater flow systems is (Zheng & Wang 1999).

$$\frac{\partial C}{\partial t} = -v\frac{\partial C}{\partial x} + D_L\frac{\partial^2 C}{\partial x^2} - \frac{\partial q}{\partial t}$$

(4.7)

where C is concentration in water (mol/kgw), t is time (s), v is pore water flow velocity (m/s), x is distance (m), D_L is the hydrodynamic dispersion coefficient [m^2/s, with D e the effective diffusion coefficient, and the dispersivity (m)], and q is concentration in the solid phase (expressed as mol/kgw in the pores).The term represents advective transport, represents dispersive transport, and the change in concentration in the solid phase due to reactions (q in the same units as C). The usual assumption is that v and D_L are equal for all solute species, so that C can be the total dissolved concentration for an element, including all redox species.

4.7.7 Procedure for Ground Water and Contaminant Transport Model

The following procedure was adopted to model the groundwater and contaminant transport

- Parameters which describe and characterize the physical framework of the study area are identified.

- Based on the collected and estimated data at the boundary of the study area the hydrological parameters are defined.

- Aquifer parameters such as hydraulic conductivity, specific storage, specific yield and porosity characterization in the specified layers of the model.

- Model calibration is done by matching the observed data with the simulated data using an appropriate technique.

- Model validation is done following the calibration.

- Sensitivity analysis is done to identify those parameters which need to be accurately estimated.

- Prediction with different scenarios to propose efficient management options.

4.7.8 Interview Schedule

In this study, the major issues which have been considered are socio economic issues arising due to the construction of tunneling for the Metro rail corridor, reduction of water level and the issues related to the quality deterioration of the water around the Metro rail corridor. In order to understand and analyze the above mentioned issues, a detailed interview schedule was designed to cover the households in and around the Metro rail Corridor. The areas are located in such a way that they are available on either side of the corridor so that it is possible to predict the exact scenario of the tunneling effect. Purposive sampling method was adopted on the basis of spatial analysis carried out in the study area.

CHAPTER 5

IMPACT ON WATER QUANTITY

5.1   GENERAL

Urbanization in any form will have both positive and negative impacts equally. The prevalent water problems like quantity shortage, environment pollution and ecological damage restrict urban sustainable development. In India, pollution and over extraction are important components of the groundwater problem. Historically, a distinct separation in the consideration of water quantity and water quality concerns has existed, with most of the attention being given to the provision of required quantities. In urban areas, generally available surface water resources are inadequate to satisfy the entire water requirements. So, the reliance on ground water has increased over the years. In most of the States in India, withdrawal of groundwater for both agricultural and industrial use has been more than what can be recharged.

5.2   TUNNELLING AND ITS EFFECT ON WATER REGIME

Metro rail construction is a massive urbanization step which involves tunneling for the underground construction. The tunneling assignment of the Chennai Metro is considered as one of the most complicated due to the complex soil structure. The water table in Chennai too is higher compared to other metropolitan cities in India. During rainy season it is almost at the ground level, hence the tunneling still gets complicated. All these factors lead to the change in water table.

Tunneling causes ground water drawdown and associated ground settlements. Tunneling beneath the ground water table causes in the state of stress and pore water pressure distribution. Hence, the assessment of water levels becomes a need for this study.

5.3   SPATIAL ANALYSIS USING GIS

Geographic Information System (GIS) is a comprehensive system of figure data management, attribute data management and spatial data analysis based on computer and it is a new margin research area that synthesizes the computer science, management science, information science, spatial science, geology science, environment science and so on. Geographical Information System (GIS) based groundwater studies focus on the preparation of hydrogeomorphological maps, interpretation of lineaments and integrated terrain analysis. Data is the core and emphasis of Geographical Information System (GIS). The Remote Sensing, Geographical Information System (GIS) and other assessment techniques have been used for a long time to study groundwater in terms of its movement, quantity and quality. It has been a new trend to introduce the GIS technique to the groundwater research and then analyze the feature and regulation of groundwater flow field by informational and visual modeling.

The spatial analysis was done for the pre monsoon and post monsoon values of the water level. The pre and post monsoon values have been separated into three phases before, during and after the construction of underground corridors. The values are given as an input along latitude and longitude in the spread sheet and it was spatially analyzed to study the effect of the corridor in the water level. It was done and it is shown in the Figure 5.1.

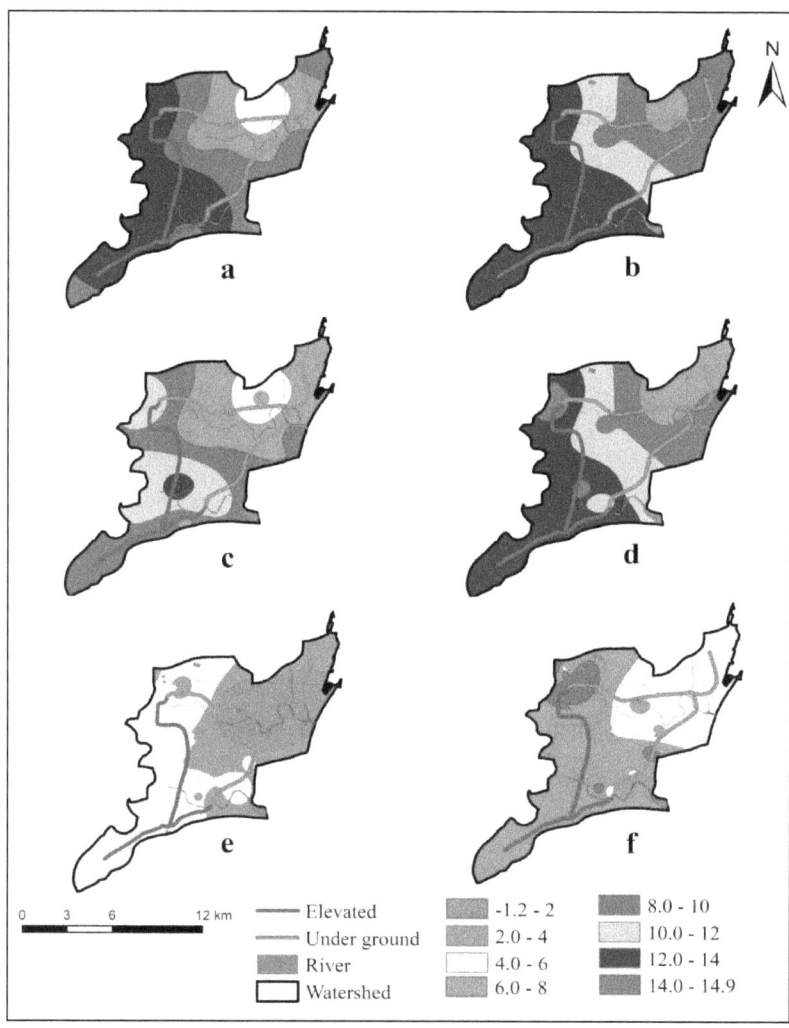

Figure 5.1   Spatial Analysis of Water level (a) Before Construction
(premonsoon) (b) Before Construction (postmonsoon)
(c) During Construction (premonsoon) (d) During
Construction (postmonsoon) (e) After Construction
(premonsoon) (f) After Construction (postmonsoon)

Before the construction, in the corridor I stretching from Central to KMC, the premonsoon water levels above mean sea level were found to be 4.0 to 6.0m. The water level of 6.0 to 8.0m exists in the stretch of KMC to Anna Nagar. From Anna Nagar to Thirumangalam, 50% of the stretch from Anna Nagar towards Thirumangalam is covered with 8.0 to 10.0m range and to the end of the Thirumangalam, the water level seems to be 12.0 to 14.0m. In the corridor II, from Central to Gemini, the water level seems to be 6.0 to 8.0m. From Gemini to Teynampet and also to the half of the stretch from Teynampet to Saidapet, the water level exists in the range of 8.0 to 10m. The remaining distance upto Saidapet is covered with the range of 12.0 to 14.0m of water level.

During the construction of the corridor stretch I, the same premonsoon values show a variation in the water level. The stretch that showed the 12.0 to 14.0m range got decreased to 8.0 to 10.0m. Similarly in the corridor II, the water level range 12.0 to 14.0m range got vanished again and it is shrink to 8.0 to 12.0m range. After the construction due to the excess withdrawal of water during the construction and also due to the aquifer properties have got change to tunneling, almost in 50% of the stretch form Central to Thirumangalam the water level got reduced to -1.2 to 2.0m. A little stretch upto Shenoi Nagar had the range from 2.0 to 4.0m and to the end of the Thirumangalam it only had the range from 4.0 to 6.0m of water level above mean sea level.

Comparing the postmonsoon values, before construction the water level in the corridor I, nearly 20% of the stretch from Central to Thirumangalam had the water level available in the range was 6.0 to 8.0m.The next 30% of the stretch was covered with 8.0 to 10.0m of water level. The water level for 10.0 to 14.0 m range is available upto the end of Thirumangalam.Similarly in the corridor II the water level range of 8.0 to

10.m was available from Central to Gemini. From Gemini 80% of the stretch upto little amount the water level seemed to be 10.0 to 12.0 m and the remaining distance was covered with 12.0 to 14.0 m range.

During the construction because of tunneling, the corridor I showed a small variation in the initial stretch with the water level standing from 8.0 to 10.0m which got reduced to 6.0m to 8.0m.In corridor II, the stretch that showed the water level 12.0 to 14.0m was reduced to 10.0 to 12.0m.After the construction the post monsoon values in corridor I, Central to Shenoi nagar the water level had reduced to 4.0m to 6.0m.Shenoy nagar to Anna Nagar it was 6.0m to 8.0m.Upto Thirumangalam 8.0m to 10.0m of water level range existed. In corridor II,Central to Gemini the water level 4.0 to 6.0m is available and from the Gemini to the end of the underground stretch the water level of 6.0 to 8.0m exists. From the spatial analysis the water level had got reduced after the construction, and the disturbance in the water level that occurred during the construction was clearly visualized. It clearly proved the fact that tunneling results in the reduction of water level.

5.4 QUANTITY ANALYSIS OF WATER

The post monsoon and pre monsoon values of water level above mean sea level were taken for comparison. Three years (2009, 2010, and 2011) of data were taken to study the water level pattern before the tunnel construction. The tunneling process fully started in a stretch from 2012 and the construction attained the completion by 2015. Hence, the period 2012-2015 was considered as during tunnel construction phase. The years 2016-2017 are considered as the data for 'after the construction'. For both before and during the construction phase secondary wells were taken. For 'after the construction phase' primary data which were taken for two years were used. Data which were separated for the above mentioned three phases are given in Table 5.1, Table 5.2 and Table 5.3.

Table 5.1 Pre-monsoon and Post-monsoon values of water levels (m) above mean sea level before tunnel construction

S.No	Location	Latitude	Longitude	Elev (m)	2009		2010		2011	
					Post Monsoon	Pre Monsoon	Post Monsoon	Pre Monsoon	Post Monsoon	Pre Monsoon
1	Tandiarpet	13.1272	80.2900	12	10.1	9.1	8.1	6.0	8.8	7.3
2	Vepery	13.0853	80.2607	12	6.8	3.4	6.8	7.3	6.9	9.9
3	Chepauk	13.0633	80.2812	13	9.2	8.5	9.2	8.4	10.4	8.7
4	Saidapet	13.0224	80.2195	15	12.4	10.0	12.7	9.1	9.4	10.2
5	Guindy	13.0102	80.2157	15	14.5	9.6	13.5	11.7	13.8	11.6
6	Aminjikarai	13.0699	80.2245	15	10.6	7.3	9.7	7.6	10.5	8.5
7	Tirumangalam	13.0835	80.1945	15	13.9	12.2	14.1	13.7	14.5	13.5
8	K.K.Nagar	13.0298	80.2130	16	14.4	12.9	15.0	12.7	14.5	12.8

Table 5.2 Pre-monsoon and Post-monsoon values of water levels (m) above mean sea level during tunnel construction

S.No	Location	Latitude	Longitude	Elev (m)	2012		2013		2014		2015	
					Post Monsoon	Pre Monsoon	Post Monsoon	Pre Monsoon	Post Monsoon	Pre Monsoon	Post Monsoon	Pre Monsoon
1	Tandiarpet	13.1272	80.2900	12	7.7	7.9	7.2	7.5	5.9	4.8	5.7	4.6
2	Vepery	13.0853	80.2607	12	7.4	4.0	5.8	3.7	6.1	4.6	6.1	4.5
3	Chepauk	13.0633	80.2812	13	10.7	7.6	9.1	8.3	9.1	8.0	9.0	7.9
4	Saidapet	13.0224	80.2195	15	13.3	10.0	12.6	11.5	10.8	8.9	9.8	8.7
5	Guindy	13.0102	80.2157	15	12.1	11.8	11.8	7.5	13.9	10.9	14.3	10.4
6	Aminjikarai	13.0699	80.2245	15	10.7	7.1	8.0	6.3	9.8	8.2	9.8	8.1
7	Tirumanglam	13.0835	80.1945	15	14.0	13.1	12.6	10.9	15.0	14.3	13.8	12.8
8	K.K.Nagar	13.0298	80.2130	16	14.5	14.2	13.9	13.6	14.8	14.0	13.6	13.4

Table 5.3 Pre-monsoon and Post-monsoon values of water levels (m) above mean sea level after tunnel construction

S.No	Location	Latitude	Longitude	Elev(m)	2016		2017	
					Post Monsoon	Pre Monsoon	Post Monsoon	Pre Monsoon
1	Bible society	13.0825	80.2782	8	4.48	1.5	4.45	1.18
2	Aziz mulk 3rd st	13.0563	80.2534	9	2.71	-0.09	2.57	-0.45
3	GAA 9th street	13.0564	80.2516	11	4.31	0.27	4.19	0.05
4	Chokkalingam nagar	13.0467	80.2523	13	7.41	1.81	7.38	1.49
5	Siva Sankaran road	13.0451	80.2502	13	10.96	6.26	10.59	5.65
6	Lotus colony (1)	13.0301	80.2401	12	4.86	4.16	4.79	3.99
7	Lotus colony (2)	13.0296	80.2407	13	7.56	4.84	7.34	4.75
8	Saidapet Jeenis road	13.0210	80.2244	12	5.34	2.25	5.19	2.26
9	Saidapet	13.0224	80.2195	13	8.96	6.76	8.75	6.74
10	Kilpauk medical college	13.0784	80.2429	12	5.36	1.41	5	1.39
11	Mc Nichols st	13.0748	80.2425	9	3.01	-1.25	2.5	-1.31
12	kathiravan colony(1)	13.0897	80.2056	13	7.46	4.76	7.4	4.55

Table 5.3 (Continued)

| S.No | Location | Latitude | Longitude | Elev(m) | 2016 | | 2017 | |
					Post Monsoon	Pre Monsoon	Post Monsoon	Pre Monsoon
13	kathiravan colony(2)	13.0894	80.2056	13	8.86	5.58	8.81	5.46
14	16st main road,anna nagar(1)	13.0902	80.2064	15	9.09	7.35	8.63	7.02
15	16st main road,anna nagar(2)	13.0897	80.2063	15	9.96	7.53	9.94	7.14
16	school of handicapped	13.0932	80.1983	14	8.6	6.4	8.7	6.3
17	vanavil appartments(1)	13.0965	80.1977	14	6.11	3.84	6.08	3.74
18	vanavil appartments(2)	13.0959	80.1978	14	5.31	2.74	5.01	2.65
19	AP flates 6th street	13.0989	80.1941	15	8.76	7.5	8.65	7.05
20	united placement colony	13.0974	80.1933	15	10.14	8.6	10.09	8.58

From the Tables shown , the minimum to maximum values of premonsoon and post monsoon water levels are plotted for the three phases and shown in Figure 5.2, Figure 5.3 and Figure 5.4.

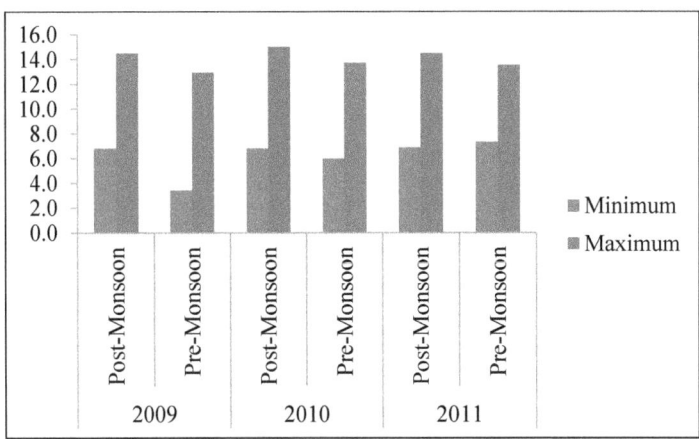

Figure 5.2 **Minimum and Maximum values of pre-monsoon and post-monsoon water levels (m) above mean sea level before tunnel construction**

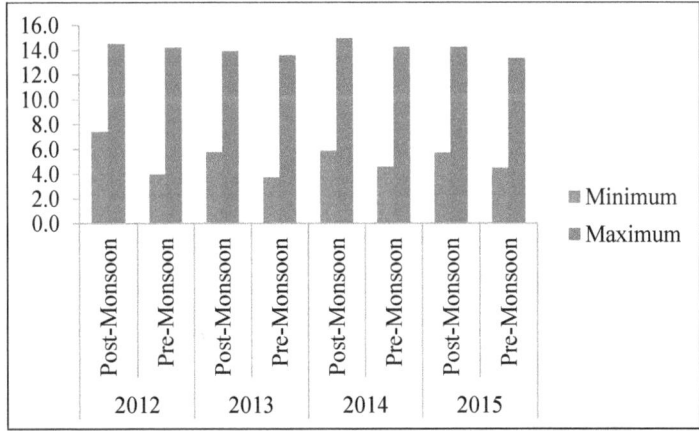

Figure 5.3 **Minimum and Maximum values of pre-monsoon and post-monsoon water levels (m) above mean sea level during tunnel construction**

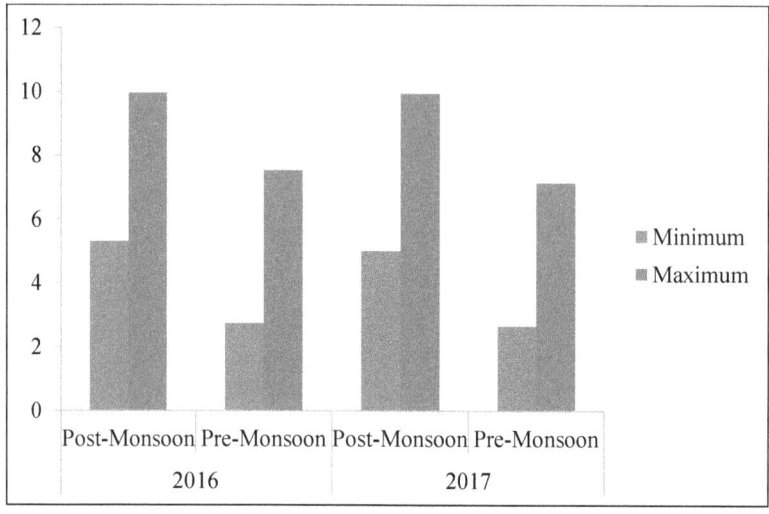

Figure 5.4   Minimum and Maximum values of pre-monsoon and post-monsoon water levels(m) above mean sea level after tunnel construction

Before the construction phase in the post monsoon season the water level above mean sea level seemed to be in the range of minimum 6.8m and to the maximum 15 m. Similarly during the pre monsoon period the water level ranged from 3.6 m to 13.7 m.

During the construction phase, post monsoon water levels were varying from the minimum of 5.8m to the maximum of 15 m. The pre monsoon water levels were ranging from 3.7m to 14m.The range of values showed a slight difference of about 1 m decrease in the water level during the post monsoon season. It could be due to the disturbance created below the ground surface by tunneling.

After the tunneling, the post monsoon values of water level showed the minimum water level as 5m and the maximum water level as 9.96m. Similarly the pre monsoon values stood at the range as 2.7m to 7.6m. This showed a major decline in the water levels. During the post monsoon season nearly 40% decline of water level had occurred from the tunneling construction phase. Similarly, in the pre monsoon, decline of water level seemed to be 59% in the maximum values range of during and after the tunnel construction.

This clearly showed that excess water had been withdrawn for the tunnel construction and also due to the excavation the properties of the aquifer had got changed and would have lost the recharge capacity. It could be also due to the sealant property of the chemicals used in the tunneling process.

5.4.1   Analysis of Opposite Wells

To observe the changes in the ground water level above mean sea level due to the tunneling effect, 20 observation wells five wells each located on either side of both corridors were selected. The pre-monsoon and post-monsoon water levels after the tunnel construction were analyzed and it is shown in Figure 5.5 and Figure 5.6. In the corridor I, Central to Thirumangalam the North side of the corridor during the post monsoon season the minimum to maximum value of water level above mean sea level seemed to be 5.26m to 8.86m. Similarly in the South side the value showed 3.01m to 10.04m.

During the pre monsoon period 1.31m and 7.5 m were the minimum and maximum water levels on the North side of the corridor. Similarly on the South side -1.25m and 8.5m seemed to be the minimum and the maximum water levels existing in the corridor. The water level ranges clearly showed that in both the pre monsoon and post monsoon periods the South side of the corridor got affected with low water levels when compared with the North side of the corridor.

In corridor II, that is from Washermenpet to Saidapet, West side of the corridor showed a decreased water level and the East side of the corridor showed an increased water level pattern. The minimum to maximum value range of water level existed in the West side of the corridor during the post monsoon period was 2.51m to 8.86m and in the East side it seemed to be in the range of 4.31m to 10.56m. During the pre monsoon season, West side of the corridor carried -0.29m and 6.66m as minimum and maximum water levels. Similarly, in the East side, 0.27m and 5.86m seemed to be the minimum and maximum water levels. In corridor I water level of nearly 1.8m is seemed to be headed up in the North side of the corridor when compared with the south side of the corridor. Similarly, a water level of about 1.5m seemed to have increased on the East side of the corridor in both the seasons.

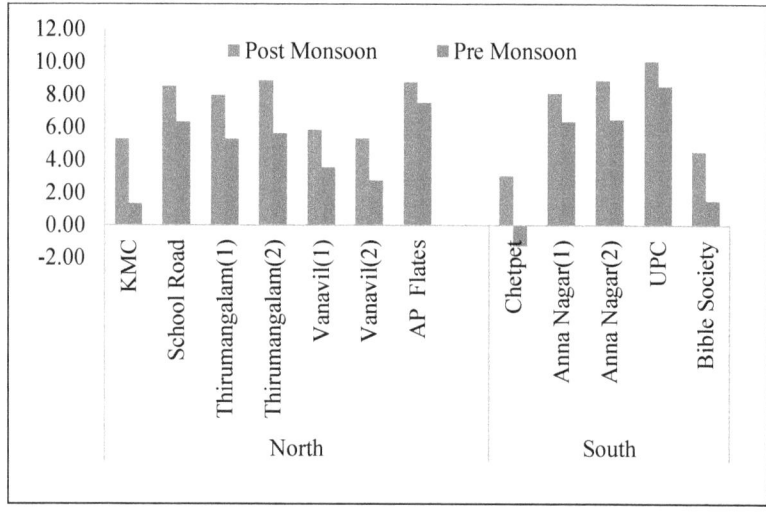

Figure 5.5 Comparison of North and South side, Pre-monsoon and Post-monsoon water levels (m) above mean sea level of corridor, Central to Thirumangalam after tunnel construction

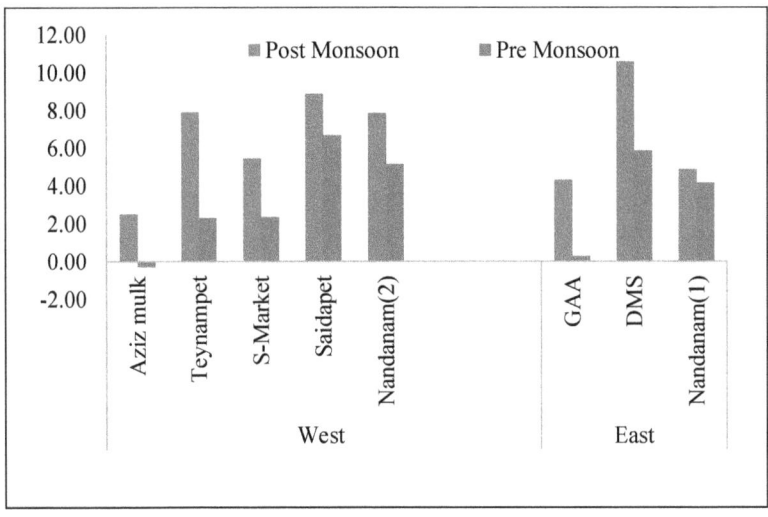

Figure 5.6   **Comparison of West and East side, Pre-monsoon and Post-monsoon water levels (m) above mean sea level of corridor, Washermenpet to Saidapet after tunnel construction**

5.5   ESTIMATION OF GROUND WATER RECHARGE

The increasing demand on groundwater leads to over-exploitation and the negative effects have attracted great attention around the world. To reduce the negative effects, it is necessary to carry out strict groundwater resources management in over-exploited areas. It can be managed by formulating different models like dynamic game model and developing some control methods. Simulating models are also carried out for the future effectiveness. Various methods can be carried out to calculate the quantity and the flow of groundwater. Quantity can be calculated by determining its inflow and outflow quantity. The inflow can be calculated by determining the amount of recharge and the location of recharge sites.

Water table fluctuation method can be used for the estimation of the recharge sites and the quantity which can be recharged or which has been recharged. The graphical mapping can also be carried out for quantifying the groundwater. Ground water volume got reduced due to the underground development around the metro rail corridors.

The application of water table fluctuation method was used to calculate the recharge quantity in terms of percentage of precipitation. The water level data which were collected for each year were listed and the mean values of the water levels were calculated. Also, for every year, the rise and fall data were calculated with the difference of water levels measured. The positive values were taken as rise and the negative values were taken as decline. Then, the net rise and net decline were calculated for each year for every observation well taken. The difference between net rise and net decline was taken as water table fluctuation. The mean annual sum of water level rises was estimated for the two phases of construction and is shown in Figure 5.7.

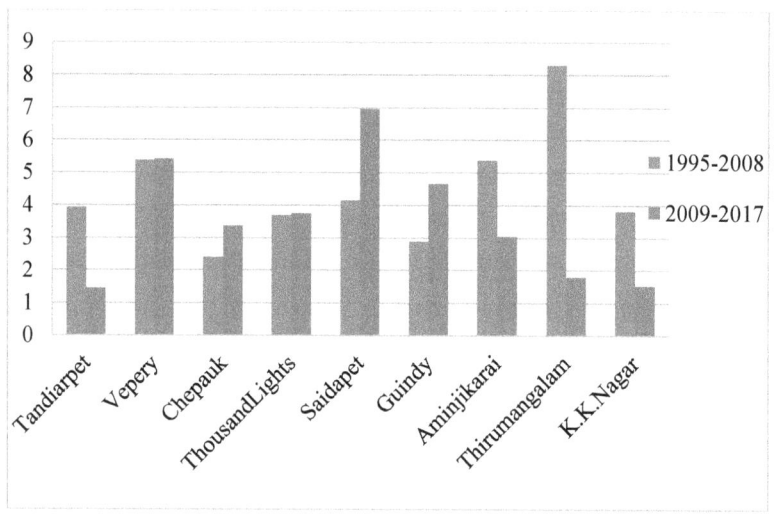

Figure 5.7   Comparison of annual sum of water level rises (m) in the Observation wells

From the above figure, it is clearly seen that the recharge pattern has been greatly affected after the construction of underground metro rail corridor. Recharge, R was estimated in terms of percentage of precipitation, %P as follows,

Recharge, R = (Net Rise/Precipitation)* S_y*100

where S_y =Specific yield of the wells concerned the specific yield of the wells collected from the lithology of the study area and the values are given below in Table 5.4.

Table 5.4 Specific yield values of the wells

Observation Well	Soil type	Specific yield
Tondiarpet	Fine Sandy	0.21
Vepery	Alluvial	0.26
Chepauk	Fine Sandy	0.21
ThousandLights	Fine Sandy	0.21
Saidapet	Coarse loamy	0.22
Aminjikarai	Alluvial	0.26
Tirumangalam	Alluvial	0.26
K.K.Nagar	Alluvial	0.26

The recharge pattern was calculated for the seven observation wells. The model calculation of recharge pattern from 1995-2017 was done for the observation well Tondiarpet and tabulated in Table 5.5. The recharge pattern of the well is shown in Figure 5.8. The recharge pattern of the remaining six wells was calculated and a graph was drawn between the precipitation and recharge and it is shown in Figure 5.9.

Table 5.5   **Computation of Recharge for the well Tondiarpet from 1995-2017**

Year	Precipitation (mm)	Mean depth(m)	Net Rise(m)	Net Decline(m)	Water table Fluctuation, h (m)	Recharge,R (%P)
1995	1525	6.54	5.37	5.65	-0.28	49
1996	1567	6.39	2.25	2.97	-0.72	30
1997	2016	6.8	3.23	3.13	0.1	34
1998	1079	6.76	0.71	3.19	-2.48	14
1999	1151	5.36	1.24	2.09	-0.85	23
2000	1080	5.7	4.61	6.09	-1.48	58
2001	1626	5.48	7.78	6.13	1.65	56
2002	1399	5.9	2.96	2.4	0.56	44
2003	739	5.26	0.3	2.68	-2.38	9
2004	1210	5.98	5.65	6.11	-0.46	65
2005	2566	3.96	9.43	6.57	2.86	77
2006	1323	7.37	1.22	2.45	-1.23	19
2007	1310	6.94	2.31	2.83	-0.52	37
2008	1596	7.25	1.15	1.45	-0.3	15
2009	1181	7.03	5.37	5.65	-0.28	62
2010	1633	6.05	1.15	2.6	-1.45	15
2011	1834	5.12	0.92	3.9	-2.98	11
2012	1053	5.47	1.43	2.23	-0.8	29
2013	1094	5.01	0.96	2.1	-1.14	18
2014	1544	5.46	0.9	0.26	0.64	12
2015	2165	3.78	1.5	0.5	1	15
2016	1090	3.5	3.78	1.35	2.43	73
2017	1681	3.57	3.4	2.11	1.29	42

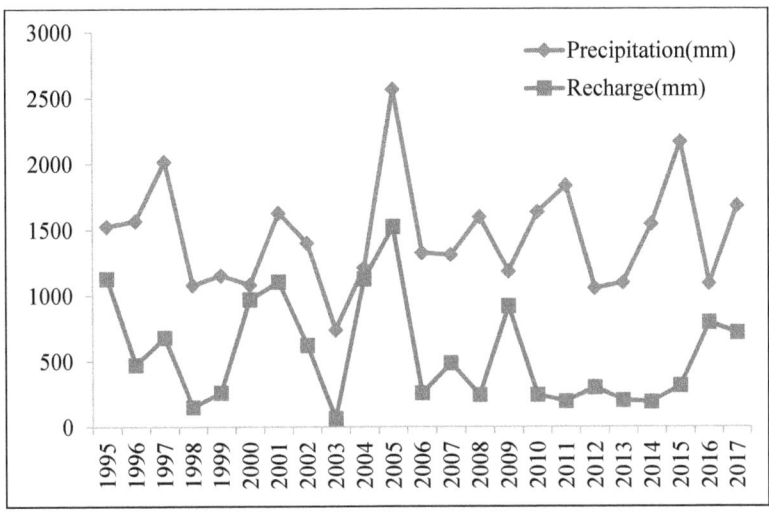

Figure 5.8 Precipitation Vs Recharge of Tondiarpet

The observation well Tondiarpet shows a major change in the recharge, that is before construction of the barrier; it shows 44% and after it has reduced to about 29%. It shows that a greater disturbance created in the aquifer affected the recharge pattern. Similarly, in the wells like Thirumangalam and K.K.Nagar the recharge percentage has got reduced. In Thirumangalam the value drops out from 73% to 60% and in K.K.Nagar it is about 57% to 50%. Wells like Vepery, Saidapet and Aminjikarai show slight increase in the recharge percentages. In Vepery an increase of about 5% and in Saidapet this increase is about 6% and only 3% in Aminjikarai wells after the tunnel construction. The well Chepauk does not show any change in the recharge pattern before and after the tunnel construction. The recharge patterns of the above wells match water rises and decline of the wells and collectively the recharge values show lower percentages in some wells and slightly higher percentages in some wells before and after the construction of metro rail corridor. Overall, there is about 6% decrease in the recharge of the wells after the barrier is created.

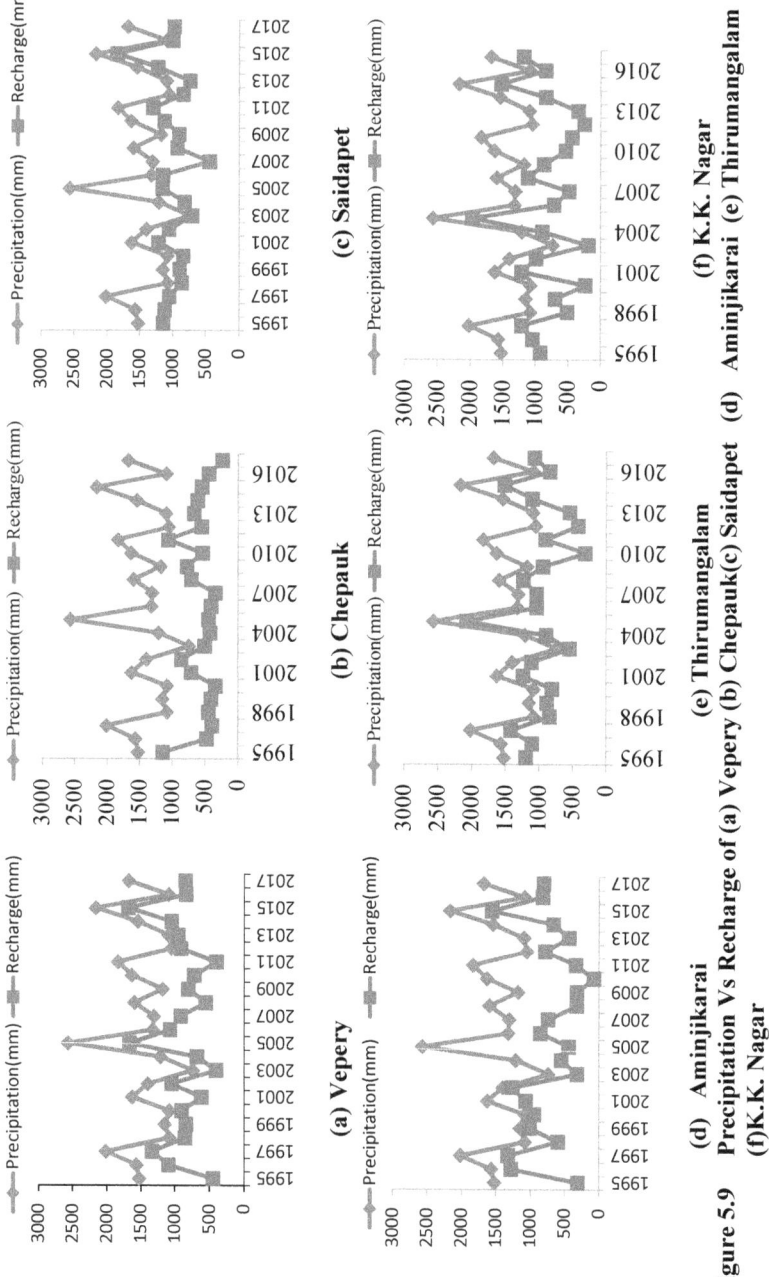

Figure 5.9 Precipitation Vs Recharge of (a) Vepery (b) Chepauk(c) Saidapet (d) Aminjikarai (e) Thirumangalam (f)K.K. Nagar

5.5.1 Estimation of Change in Ground Water Storage

Water table fluctuation method was also used to estimate the ground water storage of the study area. The volume calculation required the area of influence of the observation wells and the specific yield of the observation wells. The specific yield values were obtained from the geology map and the area of influence was computed by using Theissen polygon interpolation technique in GIS. By multiplying specific yield, Area of influence and water table fluctuation, the volume of ground water storage was obtained.

Theissen polygon interpolation was done for the observation wells and the value of area of influence was obtained in sq.km. Theissen polygon applied for the study area is shown in Figure 5.10 and the area of influence is tabulated in Table 5.6.

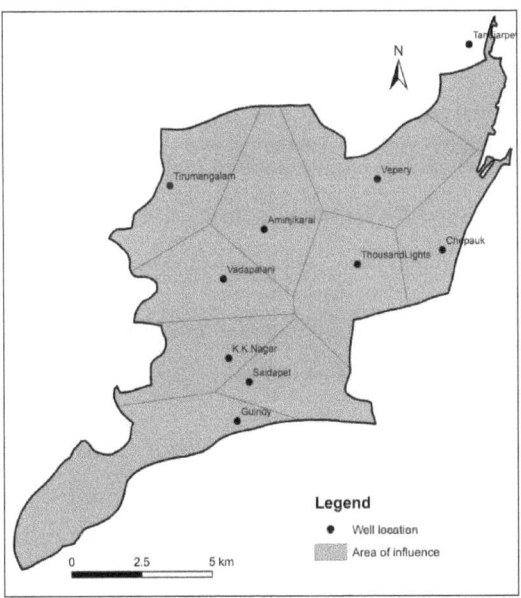

Figure 5.10 Theissen Polygon Interpolation

Table 5.6 Area of Influence of the observation wells

Well Number	Observation Well	Area_sq km
1	Aminjikarai	14.31
2	Chepauk	9.06
3	K.K.Nagar	9.62
4	Saidapet	10.44
5	Tondiarpet	5.59
6	Tirumangalam	12.41
7	Vepery	14.96
Total		76.39

The change in the ground water was calculated as,

$$\text{Change in ground water volume} = S_y \times h \times A \qquad (5.1)$$

where S_y = Specific yield (%) varies for different types of soils

h = Water table fluctuation, m

A = Area of Influence, sq-Km

Ground water storage was calculated separately for the two phases 1995-2008 and 2009-2017 separately and it is given in Table 5.7 and Table 5.8 respectively.

Table 5.7 Ground Water Storage (1995-2008)

Observation Well	1995	2008	WTF(m)	Area of Influence sq.Km	Specific yield	Storage, Km³
Tondiarpet	5.37	1.15	4.22	5.59	0.21	0.005
Vepery	1.73	2.13	-0.4	14.96	0.26	-0.002
Chepauk	8.46	2.01	6.45	9.06	0.21	0.012
Saidapet	7.22	5.55	1.67	10.44	0.22	0.004
Aminjikarai	1.18	1.25	-0.07	14.31	0.26	0.000
Tirumangalam	7.6	5.16	2.44	12.41	0.26	0.008
K.K.Nagar	2.26	1.85	0.41	9.62	0.26	0.001
					Total	**0.028**

Table 5.8 Ground Water Storage (2009-2017)

Observation Well	2009	2017	WTF (m)	Area of Influence sq.Km	Specific yield	Storage, Km³
Tondiarpet	1.15	3.78	-2.63	5.59	0.21	-0.003
Vepery	4.96	6.48	-1.52	14.96	0.26	-0.006
Chepauk	3.67	2.59	1.08	9.06	0.21	0.002
Saidapet	8.15	8.4	-0.25	10.44	0.22	-0.001
Aminjikarai	1.25	14.12	-12.87	14.31	0.26	-0.048
Tirumangalam	1.21	13.82	-12.61	12.41	0.26	-0.041
K.K.Nagar	2.05	14.75	-12.7	9.62	0.26	-0.032
					Total	**-0.128**

From the above Tables 5.7 and 5.8, the ground water storage value of the phase 1995-2008 shows a positive value and the phase 2009-2017 shows the negative; thus the scenario cleanly depicts the status of the ground water volume. Thus the water table fluctuation method proves to be very efficient in finding the ground water storage and the underground developments decrease the water level and the quantity.

5.6 GROUND WATER FLOW MODELING

Ground water flow modeling was done with the necessary inputs such as model geometry, initial heads of the observation wells, hydraulic properties, lithological data, specific storage, layer classifications and their permeability values. Water table levels and the pumping data of the wells were also given for the analysis. The flow modeling was done to predict the changes in the direction of flow due to the construction of tunneling and also the changes in the water table levels.

5.7 MODEL GEOMETRY

In the present study ,Visual MODFLOW was used to determine the groundwater flow. The line map of the Chennai metro rail corridor was given as a base map for the model input. Geo-referencing was done for two points, one for the airport as (x=1442346.432, y=8898966.237) and another for Washermenpet as (x=1457656.329, y=8910899.292) as model co-ordinates. The model simulates the ground water flow for the study area with 60 rows and 80 columns of rectangular grid pattern modeling domain with two vertical layers. The cell size of the domain is 100 m X 80 m. The input of the grid map is shown in Figure 5.11.

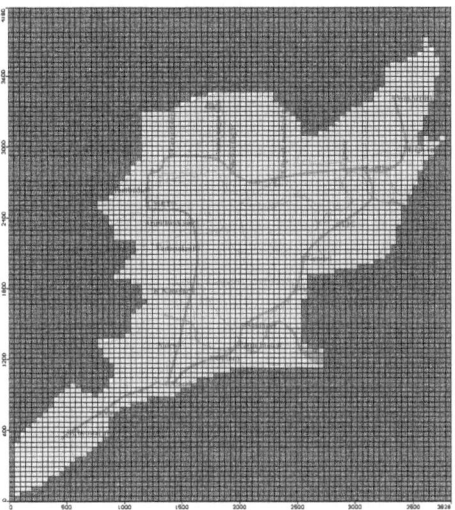

Figure 5.11 Input of the Grid map

The sources and sink were absent in the north, south and western boundary of the study area, transportation of the flow in these boundaries made them as flow boundary. The coastal area in the Eastern side of the study area was considered as constant head boundary. The constant hydraulic head of 1.0 m along the eastern boundary was taken. The flow boundary along the northern metro rail corridor area was considered as general head boundary and the Buckingham canal was assigned for river boundary. The spatial discretization of the study area with boundary conditions is shown in Figure 5.12.

Two layers were considered for modeling purpose. Clayey sand was considered as the top most layer followed by sandy layer shown in Figure 5.13 The initial condition was taken as the water level data for the year 1995 as shown in Table 5.9 and the contour map of the initial head is shown in Figure 5.14.Directions with 5.0 m, 8.0 m along North West, 10.0 m along south west and 4.0 m along southern direction were considered.

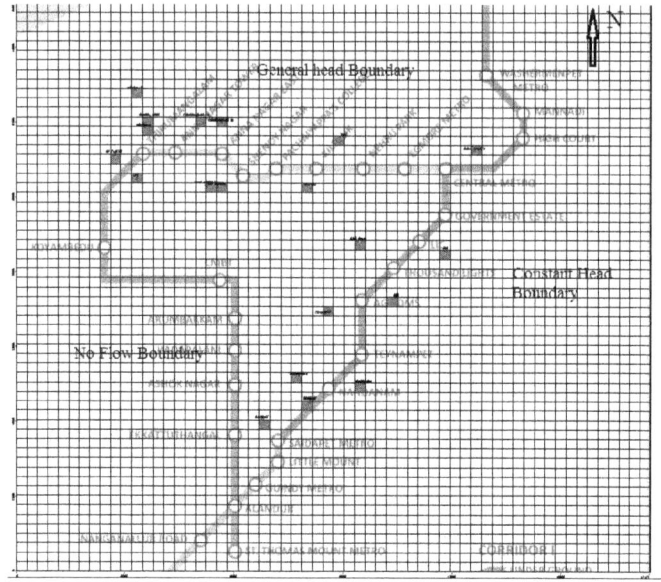

Figure 5.12 Boundary conditions in the study area

Table 5.9 Initial Head for the Observation Wells (m)

Location	X Coordinate	Y Coordinate	Initial Head(m)
Tandiarpet	14737	18865	7
Vepery	10648	14930	9
Chepauk	14969	13310	9
ThousandLights	11882	12307	6
Saidapet	8757	9143	10
Aminjikarai	8333	14120	10
Tirumangalam	4977	15316	10
K.K.Nagar	5864	9568	11
Guindy	8024	6751	8

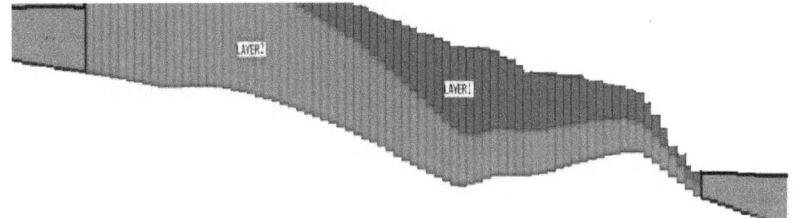

Figure 5.13 Layer Classifications in the Model

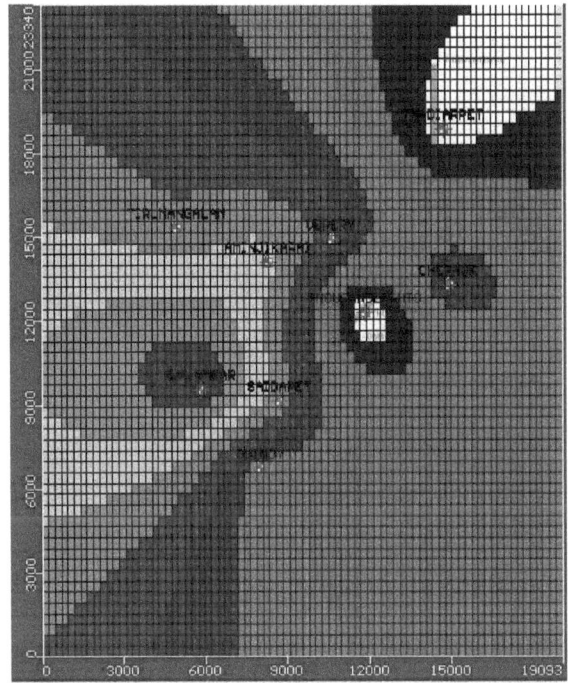

Figure 5.14 Contour Map of the Initial Head (m)

To calibrate and validate the model the monthly water level data of the observation wells for a period of 23 years were used .The water level data from June1995 to May 2005 for a period of 10 years and the data from June 2006 to May 2009 for a period of 3 years were used for model validation.

5.8   HYDRO GEOLOGICAL SYSTEM

Based on the lithological data collected from the Public Works Department (PWD) and the data collected from the nearby residents were used to generate the physical framework of the sub surface environment of the study area. The control points elevation was taken with respect to the MSL for the modeling of the aquifer. The geological cross section along the corridor from West - East and North - South direction is shown in Figures 5.15 (a) and (b).

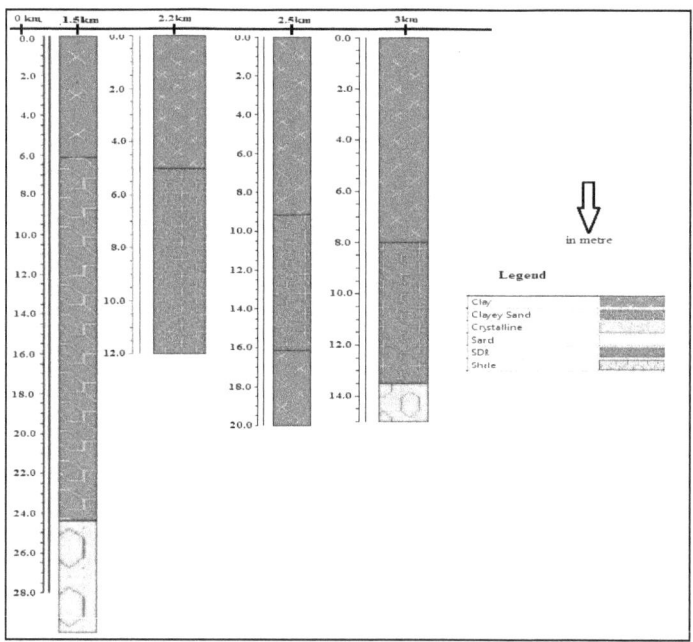

Figure 5.15 (a) Geological cross section along West to East direction

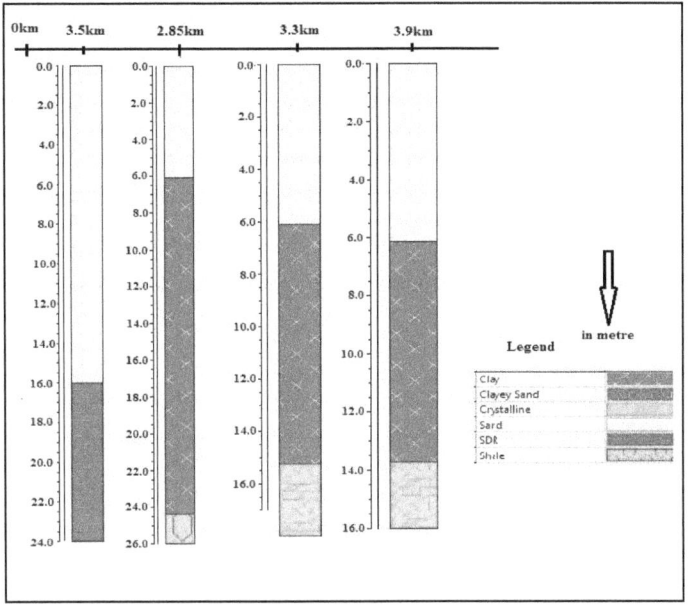

Figure 5.15 (b) Geological cross section along North to South direction

The Figures 5.15 (a) and (b) show that the corridor stretch has got separated for a fragmental distance from Central to Tirumangalam for a total distance of 9.2 km. At a distance of 1.5 km the bed rock lay at a distance of 28 m bgl. Upto 6m clay layer existed and for a depth from 6m to 24 m clayey sand and from that point for a depth of 2 m shale existed. For a 2.2 km stretch from 1.5 km the bed rock was encountered at a depth of 12 m itself and for a depth of 4.5 m clay soil and from 4.5 m to 12 m clayey sand was present.

Again for a distance of 2.5 km the bed rock lay at a distance of 20m and to a depth upto 8.5 m it was clay, 8.5 m to 16m clayey sand and upto 20 m again clay sand was present. And to the end of the stretch of 3 km, the bed rock was upto 14 m bgl. To a depth of 8m clay soil followed by clayey sand for a depth of 4m and for the rest of 2m shale existed.

The corridor stretch from Washermenpet to Saidapet covering a total distance of 13.55 km got separated for a distance 0 m – 3.5 km, 3.5 km - 6.35 km, 6.35 km – 9.65 km and upto 13.55 km. At a distance of 3.5 km the bed rock lay at a distance of 24 m bgl, upto 16m Sand layer existed and for a depth from 16m to 24 m presence of Soft Disintegrated Rock was indicated. For a 2.85 km stretch the bed rock was encountered at a depth of 26 m and for a depth of 6 m sandy soil and from 6 m to 24 m clay soil was present and further below for a depth of 2 m it was the presence of shale.

Again for a distance of 3.3 km the bed rock lay at a distance of 16m to a depth upto 6 m it was sand, 6 m to 14 m clayey sand and upto 16 m the shale soil was present. And to the end of the stretch of 3.9 km, the bed rock was upto 16 m bgl. To a depth of 6m it was sandy soil followed by clay for a depth of 8m and for the rest of 2m shale existed.

5.9 HYDRAULIC PROPERTIES

The aquifer properties such as hydraulic conductivity, storage coefficient, porosity and specific yield were assigned to the different layers of the model based on the slug test conducted in the bore hole and pumping test conducted by government agencies (PWD 2000). In addition, the values were derived from the literature (Fetter 1994, Senthilkumar & Elango 2004) and used.

The horizontal hydraulic conductivity ($K_x = K_y$) for the top soil was taken as 4 × 10-2 cm/s, 5 × 10-2 cm/s for sandy layer, 9 × 10-3 cm/s for sandy clay and 4.6 × 10-3 cm/s for weathered rocky stratum. The vertical conductivity (K_z) was assumed to be 1/10th of the horizontal hydraulic conductivity. The specific storage ranged between 1.0×10^{-4} and 2.0×10^{-3} m^{-1}. The specific yield varied from 0.07 to 0.24 and the porosity varied between 0.23 to 0.4.The initial head data were assumed as the water level data of the observation wells for the year 1996.

5.10   VISUAL MOD FLOW

MODFLOW uses the finite difference technique for the governing flow equations solutions. It simulates the three-dimensional groundwater flow and the behavior of groundwater flow systems under artificial stress conditions. The region of the flow is divided into blocks and the layers of varying thickness. For each time step the flow rate and the cumulative – volume balances can be found out for the inflow and outflow. Hydraulic conductivities may differ for each layer spatially and the storage coefficient may be heterogeneous.

5.10.1   Modflow Input

5.10.1.1   Base map

The line map of the Chennai metro rail corridor was given as a base map for the model input. Geo-referencing was done for two points one for the airport as (x=1442346.432,y=8898966.237) and another for Washermenpet as (x=1457656.329,y=8910899.292) as model co-ordinates. The model simulated the ground water flow for the study area with 60 rows and 80 columns, with two vertical layers.

5.10.1.2   Governing equations and groundwater model selection

Three-dimensional groundwater flow can be mathematically represented given the following equation (1) based on water mass balance and Darcy's law equations (Bear 1972):

$$\frac{\partial}{\partial x}\left(Kxx\frac{\partial h}{\partial x}\right) + \frac{\partial}{\partial y}\left(Kyy\frac{\partial h}{\partial y}\right) + \frac{\partial}{\partial z}\left(Kzz\frac{\partial h}{\partial z}\right) - Q = Ss\frac{\partial h}{\partial t}$$

(5.2)

K_{xx}, K_{yy}, K_{zz} are hydraulic conductivity along the x, y, z axes which are assumed to be parallel to the major axes of hydraulic conductivity;

h   = piezometric head;

Q   = Volumetric flux per unit volume representing source/sink terms;

Ss   = Specific storage coefficient defined as the volume of water released from Storage per unit change in head per unit volume of porous material.

5.10.1.3 Model conceptualization

A systematic way of describing the field conditions for the ground water flow is explained in Model Conceptualization. The conceptual model was developed based on the studies done on the water level fluctuation, bore hole lithology and geology of the study area. Since the ground water was found to occur in weathered rocks, the aquifer was assumed to be consisting of the top soil and the lower weathered and fractured rocks as one layer below the bottom as another layer, unconfined aquifer.

5.10.1.4 Pumping wells

Based on the land use of the study area and the pumping data from the water board, well inventory and their pumping rate and their usage were calculated. The pumping wells were located in the grid and the pumping rate was entered as shown in Figure 5.16. Positive rates were used for injection and negative rates were used for pumping. In the study area there were no injection wells.

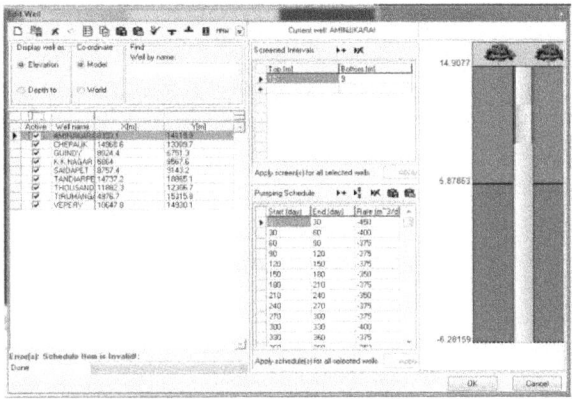

Figure 5.16 Input of Pumping Wells

5.10.1.5 Head observation wells

Excel spread sheets were created for the head observation wells for the nine secondary wells and the twenty primary wells with latitude and longitude details and observed head of the wells with the respective times. The observed head data for the secondary wells from 1995 – 2016 and the secondary wells from June 2014 to December 2016 were generated and were given as input to the model head observation as shown in Figure 5.17.

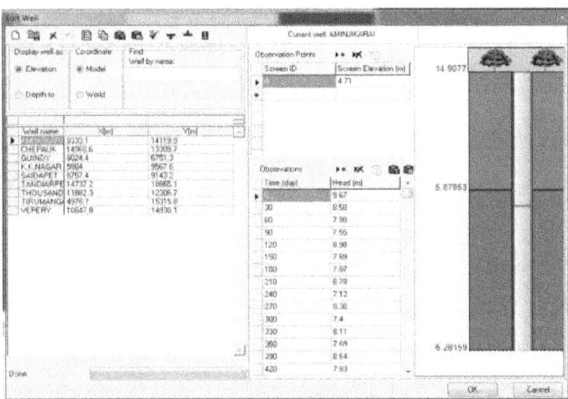

Figure 5.17 Input of Head Observation Wells

5.10.1.6   Model design

The model design included the parameters that are given in the Table 5.10 below. The input parameters given under were used to develop the calibrated model.

Table 5.10 Properties of the Aquifer

S.No	Model Property	Value
1	Hydraulic Conductivity in longitudinal direction, Kx	5.00E-05
2	Hydraulic Conductivity in lateral direction, Ky	5.00E-05
3	Hydraulic Conductivity in Vertical direction, Kz	5.00E-07
4	Specific yield, Sy	2.00E-01
5	Effective porosity	1.50E-01
6	Total porosity	3.00E-01
7	Specific storage	1.00E-05

For the model properties assigned the respective layers of permeability diagrams are shown in Figure 5.18 (a) and (b).

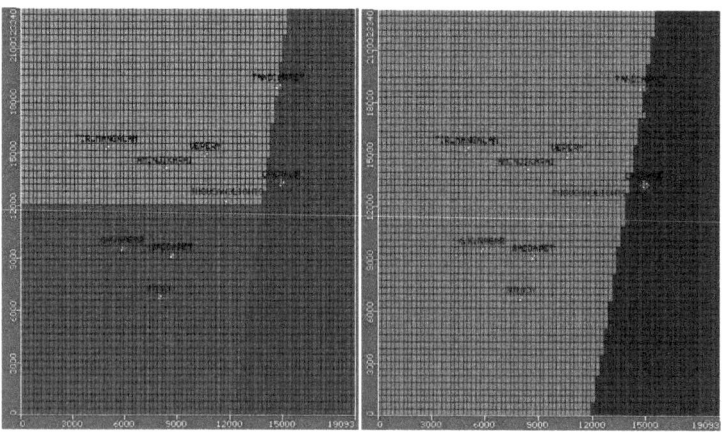

Figure 5.18 (a) Layer 1 Permeability (b) Layer 2 Permeability

5.10.2   Hydrologic Stresses

5.10.2.1   Model calibration

The model was calibrated initially for the steady state condition in order to minimize the variation between the observed and calculated head in the observation wells. Recharge values and the hydraulic conductivity values were subjected to iterative process, so that a small fraction of difference existed in the residual between the highest and lowest heads.

The transient state condition of the model was also calibrated. The calibration was carried for the time period from 1995 to 2005 (3650 days) and the correspondence of the calculated and the observed groundwater head is shown in Figures 5.19 (a) and (b). To obtain reasonably good correlation between the calculated and the observed water levels, the hydraulic conductivity and the recharge parameters were varied between the calculated and the observed water levels. From the graph of the calculated head vs the observed head for 120 days and 1440 days, it can be observed that the calculated head matches the observed head at 99 % confidence level.

The calibration residual histogram shown in Figure 5.20 indicates the normal distribution. Some of the calibrated head for the period of 10 years of the Eight observation wells are given in Figures 5.21 (a), (b), (c), (d). The calculated heads for the observation wells show fairly good agreement with observed values.

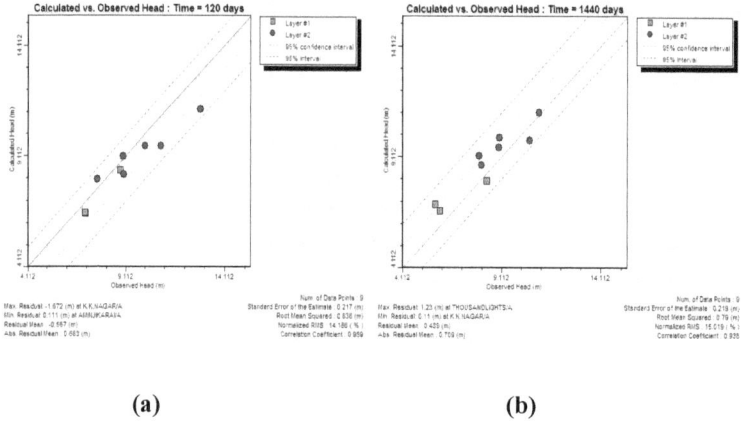

(a)   **(b)**

Figure 5.19 Calculated vs Observed Heads for (a)120 days(b) 1440 days

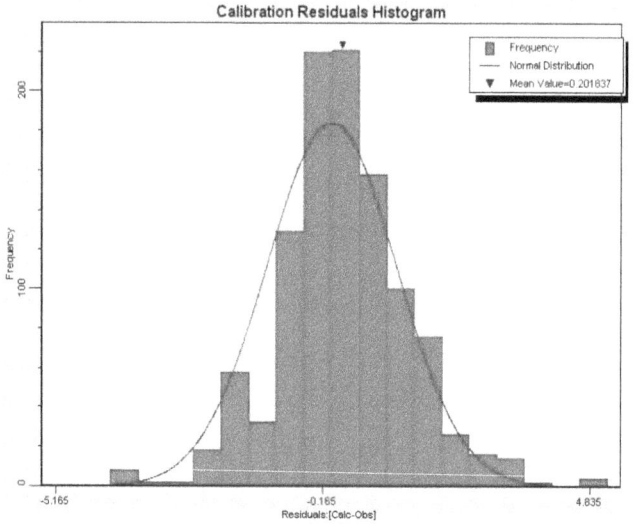

Figure 5.20 Calibrated residual histogram of Calculated vs ObservedHead

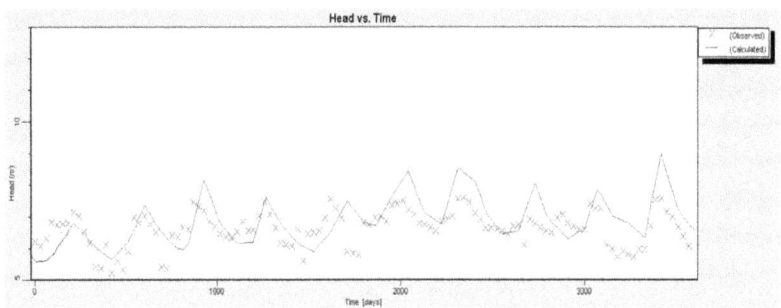

(a) Tondiarpet

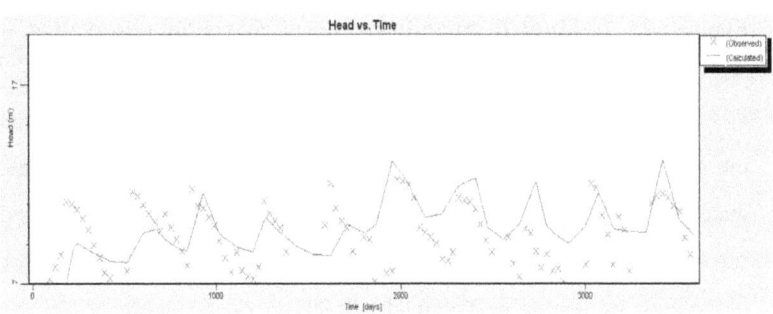

(b) Vepery

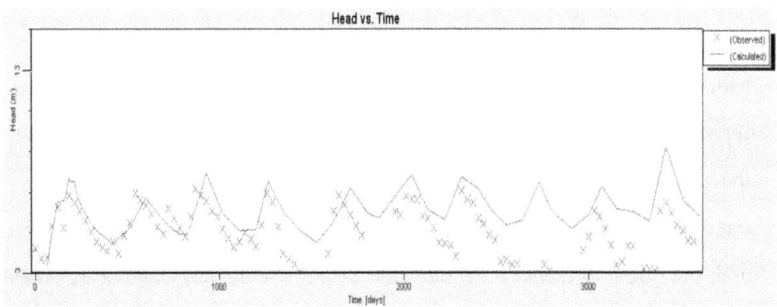

(c) Saidapet

Figure 5.21 (Continued)

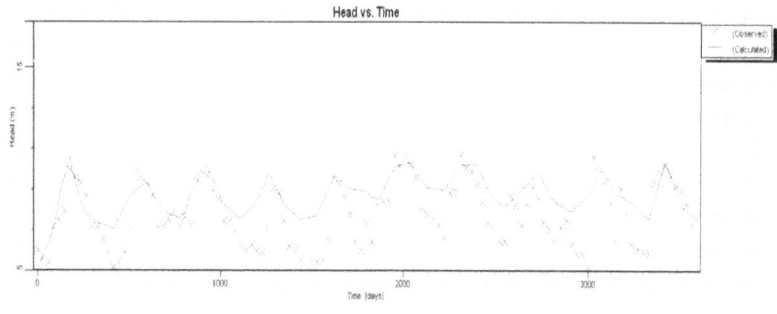

(d) Tirumangalam

**Figure 5.21 Calibrated groundwater head charts for observation wells
(a) Tondiarpet (b) Vepery (c) Saidapet(d) Tirumangalam**

The study areas field condition characterization is very essential for the good calibration. The model input parameters values were changed to match the field conditions to an acceptable limit. Model calibration was done for the period from 1995-2005. The observed and the calculated head of the observation wells were well correlated at 99% of confidence level.

5.10.2.2 Model validation

Model Validation is the process of further calibration of the model. Model validation was done for the period from 2006-2011.The model has reproduced the same values as in model calibration. Hence it is evident that the model validation responded well for the measured changes in the field conditions. Again the observed and the calculated head were well correlated and the same presented for the time step of 4140 days and it is shown in Figure 5.22.

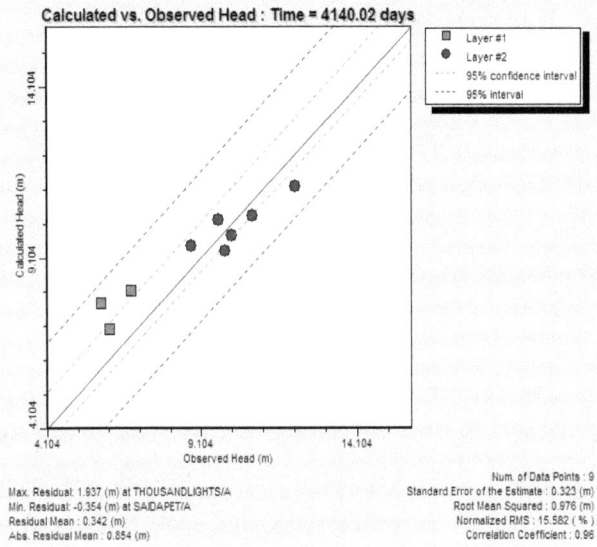

Figure 5.22 Calculated vs Observed Head for 4140 days

5.10.2.3   Model prediction

Model prediction predicts the future ground water flow and evaluates the alternatives. The outcomes of the model prediction should correlate with the assumptions in the model and also the uncertainty of the model input parameters. In this model, the prediction is done purely based on the progress of the tunneling work. The progress chart of tunneling work is given under.

Under four scenarios of progress of tunneling that is from January 2012 to December 2013, January 2014 to December 2015, January 2016 to March 2017 and the future prediction from 2017-2020 prediction was done for ground water flow analysis. In the base map the progress of tunneling for the respective periods was done by constructing the wall provision in the corridors.

5.10.2.4   Wall constructed for the phase 2012-2013

During the phase of construction of tunneling, the progress of tunneling was 0.81 Km from Shenoy Nagar to Tirumangalam for the total stretch of 5.594 Km. Out of 6.776 Km from Egmore to Shenoy Nagar, 1.528 Km stretch was completed. For the total of 5.736 Km, Saidapet to AGDMS, the stretch of 1.912 Km was completed during the phase of construction of 2012-2013.

The distance of 1.205Km was completed in the total stretch of 3.616 Km from May Day Park to AGDMS. The total stretch from May Day Park to Central is 2.080 Km, out of which 0.193 Km was completed during this period. From Washermenpet to Egmore, 1.465 Km was completed in the total stretch of 4.445Km.The wall construction for the phase of 2012-2013 is shown in Figure 5.23.

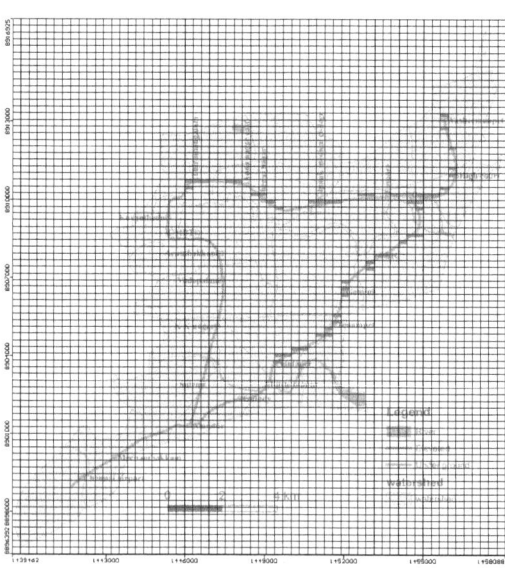

Figure 5.23 Wall Construction for the phase 2012-2013

5.10.2.5   Wall constructed for the phase 2015-2016

During the phase of construction of tunneling, the progress of tunneling was 2.797 Km from Shenoy Nagar to Thirumangalam for the total stretch of 5.594 Km. Out of 6.776 Km from Egmore to Shenoy Nagar, 2.429 Km stretch was completed.For the total of 5.736 Km, Saidapet to AGDMS, the stretch of 1.725 Km was completed during the phase of construction of 2015-2016.

The distance of 1.292 Km was completed in the total stretch of 3.616 Km from May Day Park to AGDMS. The total stretch from May Day Park to Central is 2.080 Km, out of which 0.835 Km was completed during this period. From Washermenpet to Egmore, 1.525 Km was completed in the total stretch of 4.445 Km. The wall construction for the phase of 2015-2016 is shown in Figure 5.24.

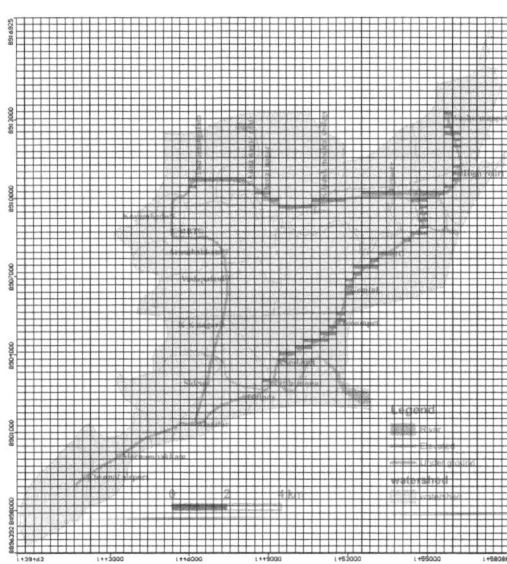

Figure 5.24 Wall Construction for the phase 2015-2016

5.10.2.6   Wall constructed for the phase 2017-2020

During the   phase of construction of tunneling, the progress of tunneling was 1.987 Km from Shenoy Nagar to Thirumangalam for the total stretch of 5.594 Km. Out of 6.776 Km from Egmore to Shenoy Nagar, 2.819 Km stretch was completed. For the total of 5.736 Km, Saidapet to AGDMS, the stretch of 2.099 Km was completed during the phase of construction of 2017-2020. The distance of 0.205 Km was completed in the total stretch of 3.616 Km from May Day Park to AGDMS. The total stretch from May Day Park to Central is 2.080 Km, out of which 1.052 Km was completed during this period. From Washermenpet to Egmore, 1.445 Km was completed in the total stretch of 4.445 Km. The two corridors were almost completed at the end of 2017 . Hence the wall was fully constructed in the third phase of 2017 -2020. The wall construction for the phase of 2017-2020 is shown in Figure 5.25.

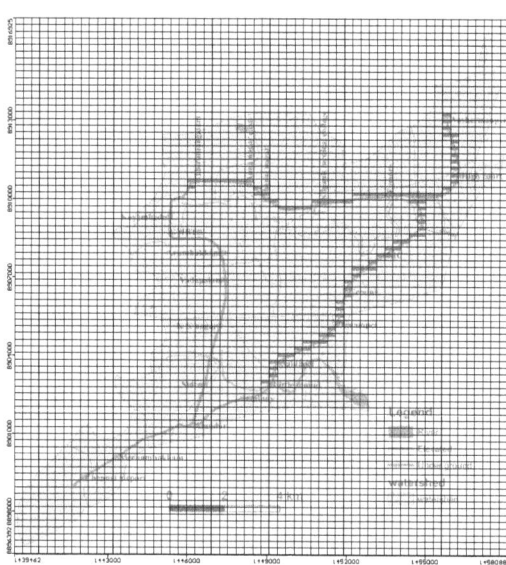

Figure 5.25 Wall Construction for the phase 2017-2020

5.10.2.7 Velocity direction

Under the three scenarios of the progress of tunneling the velocity direction seemed to follow the similar pattern but after the complete construction of the tunneling the flow direction changed. Before the construction the contour gradients which were closely spaced got distorted after the construction of the tunnel. The Head equipotentials before the tunneling varied from 6.5 m to 11.5 m, but after the construction the head equipotentials varied from 6.00 to 15.00.Hence it became evident that the tunnel construction behaves as a barrier with the changes in the direction of the flow making the head equipotentials higher at one end and lower at the opposite side of the tunnel. Velocity direction during the calibration of the modeling and the velocity direction predicted after the full construction tunneling work are shown in the Figure 5.26 (a) and (b). The column and the row views of the velocity direction of the corridors are shown in the Figure 5.27.

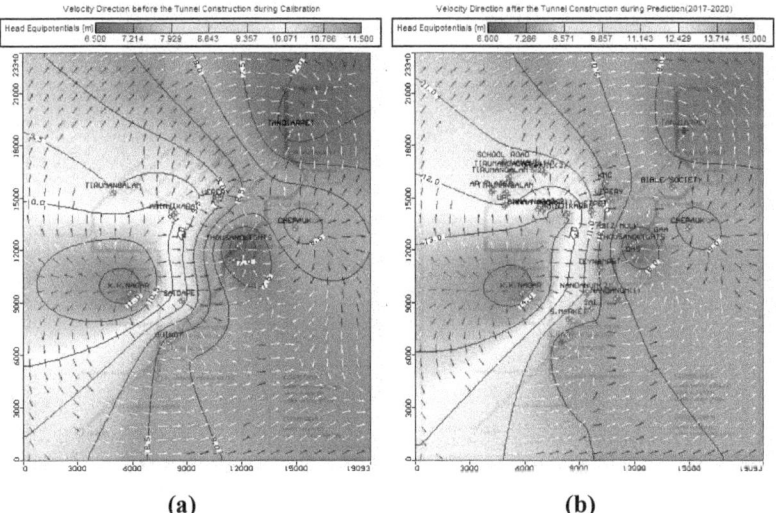

(a)   (b)

Figure 5.26 (a) Velocity Direction during the Calibration (b) Velocity direction after the full construction of tuneling

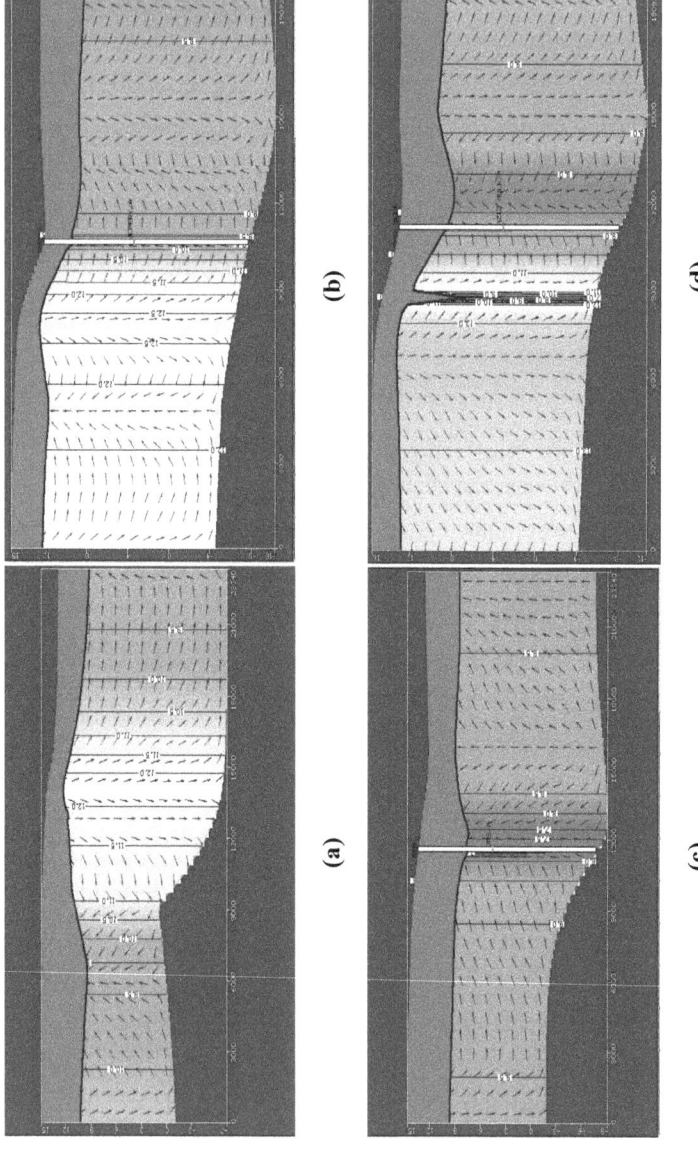

(a)

(b)

(c)

(d)

Figure 5.27 **Column View and Row view of the corridors I and II (a)Column view of corridor I(b)Row view of corridor I (c) Column view of corridorII (d)Row view of corridor II**

5.10.2.8 Equipotential heads

The Equipotential heads from the model are shown in Figure 5.28. The equipotential head for the time step 3700 days, that is the period before the construction showed water level below the ground level and was found to be 8.0 m to 10.5 m.

The time step of 7920 days that is during the construction showed the water level below the ground level and was found to be in the range of 8.5 m to 11.5m.This indicated a drawdown of 0.5m to 1m around the study area. This might have occurred due to the excess withdrawal of water during the construction and also aquifer disturbance caused in the aquifer.

The time step of 9920 days that after the construction of tunnel showed the water level below ground level to be the range of 9.0m to 14 m. This showed a major drawdown in the water level of nearly 2.5m.The results clearly showed that the water level pattern and the recharge of the wells we disturbed in a larger manner.

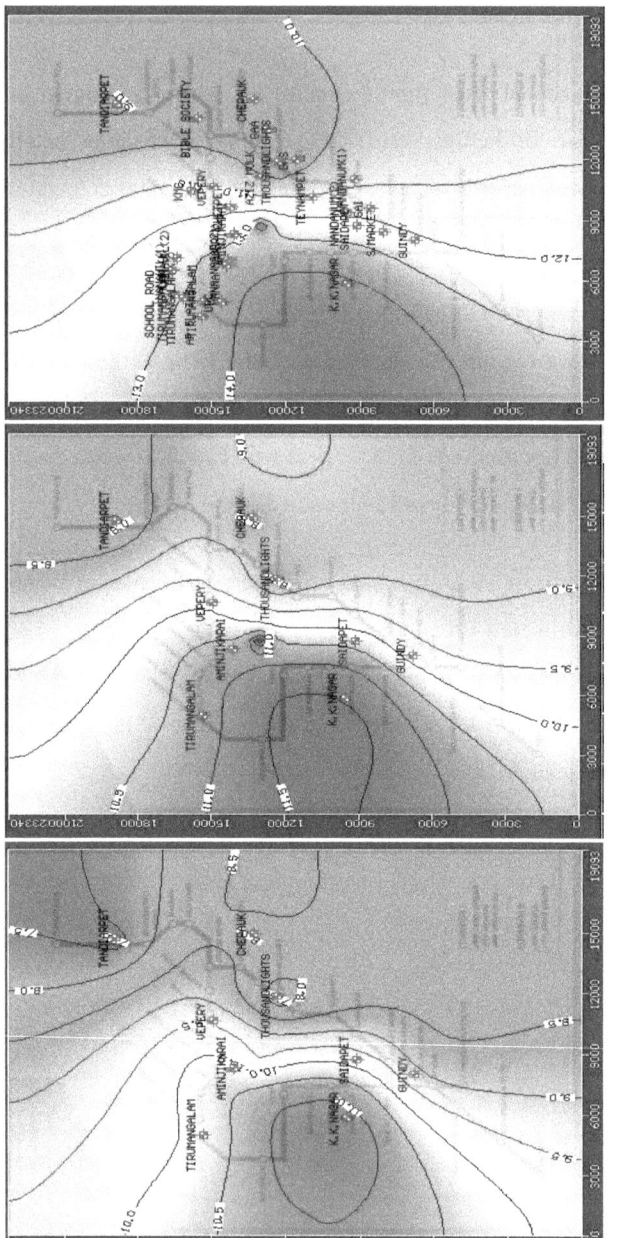

Figure 5.28 Head Equipotentials for the time steps (a) 3700 days (b)7920 days (c) 9920 days

5.10.2.9 Computation of storage from the model

The cumulative storage of the different time steps are shown in Table 5.11. The storage before the construction was found to be 5335368 m³, during the construction it was found to be 942532 m³ and after the construction it stood at 1121142 m³.The storage during the construction was much reduced from the storage before the construction. Once the disturbance got stabilized and the recharge supported the aquifer, the storage after the construction was found to be higher than that during the construction but it was very much lower than before the construction. It clearly depicted that the storage of the aquifer also very much affected after the construction of these corridors. The well storage also got much reduced from before the construction of the corridors. The mass outputs of the respective time steps in the Table 5.11 are shown in Figure 5.29. Mass balance graph is shown in Figure 5.30.

Table 5.11 Cumulative storage, Recharge and Well storage of the different time steps

S.No	Time step(days)	Storage(m3)	Recharge(m3)	Well Storage(m3)
1	360	5314899	5586587	1051380
2	5400	10650266	77630968	6419520
3	5760	10810806	83610104	6704067
4	7920	11753338	117425800	8616464
5	8280	11893535	123080488	8812664
6	9920	13014677	133833248	9368109

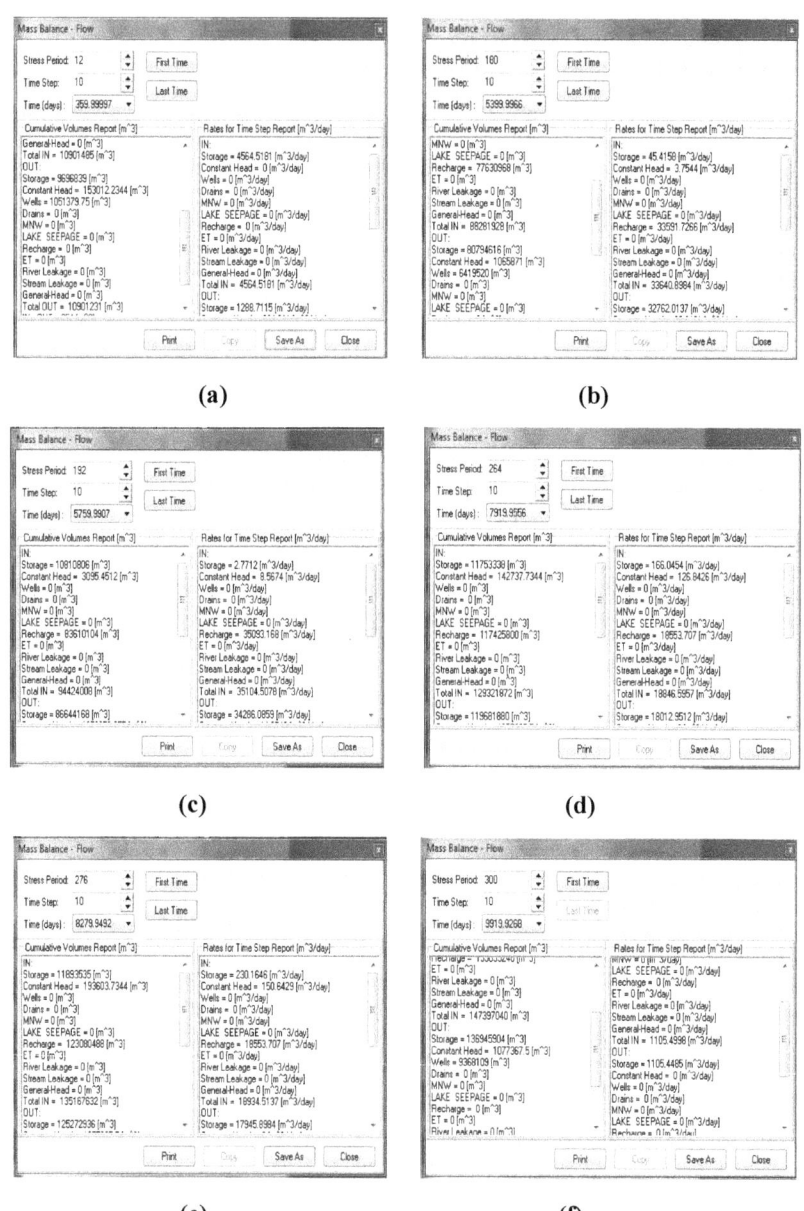

(a)   (b)

(c)   (d)

(e)   (f)

**Figure 5.29   Mass Balance outputs (a) 360 days (b) 5400days (c) 5760days
(d) 7920days (e) 8280days (f) 9920days**

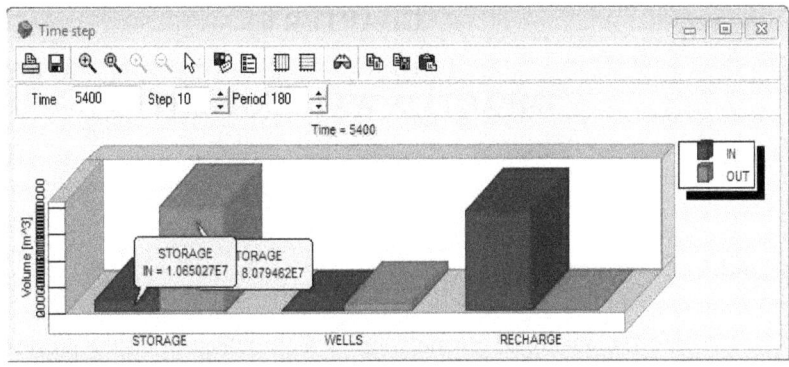

(a)

(b)

Figure 5.30 Mass Balance Graph for the time step of (a) 5400 days (b) 8280 days

5.11 SUMMARY

This chapter summarizes the effect on water quantity on either side of the corridor I and corridor II through spatial analysis. The modeling was done by constructing walls for the tunnels and predicted the changes in the flow direction and also the mass balance to find the storage in the aquifer.

CHAPTER 6

IMPACT ON WATER QUALITY

6.1   GENERAL

The quality of ground water is the resultant of all the processes and reactions from the time it condenses in the atmosphere to the time it is discharged by a well. Therefore, the quality of ground water varies with place, water table depth, and from season to season and is primarily dependent on the dissolved solids present in it. The sub surface development also impacts the ground water quality and has adverse effects on it. The various types of Sub-structures are underground drainage, water lines and the depth not more than 4m, but the sub-structure development like tunneling breaks and changes the geological formations, water level and water quality.

As people of Chennai mainly depend on bore well water for their domestic and drinking purposes, water quality assessment becomes imperative and plays a vital role. Due to growing urbanization and industrialization, the environment is facing various threats like river pollution, soil degradation and polluted air quality. The parameters are greatly affected due to the chemicals used in the tunneling methodology. The chemicals which are used in the tunneling process and the prescribed limits of the water quality parameters are shown in the Table 6.1 and Table 6.2.

Table 6.1 Chemicals used in Tunneling process

S.No	Chemical	Application
1	Tam Soil 190 CF	Polymer gives the cohesive characteristic 100% soluble in water, carries pH value of 6.5 -8.5,Soil conditioning foam.
2	Tam Soil 800 AD	Anti dust, Anti wear
3	Tam Soil 600 CP	Polymer
4	Tam Seal TG11	Tail Sealants
5	Tam Grease BS1	Main Bearing Lobricant
6	TamCem 8BFG	Grouting
7	Collaidal Silica	Mineral used as Injection chemical
8	Poly urea resins	Injection chemical

Table 6.2 Prescribed limits of Water quality parameters

S.No	Water Quality parameter	Most desirable limit	Max allowable limit
1	pH	6.5-8.5	9.2
2	TDS(mg/l)	500	1500
3	TH(mg/l)	80-100	500
4	Cl(mg/l)	200	600
5	Fl(mg/l)	0.6-1.2	1,5

6.2 QUALITY ANALYSIS OF WATER

The water quality parameters such as pH, Total dissolved solids, Total Hardness, Chloride and Fluoride were collected and separated into three phases before, during and after the tunnel construction. The post monsoon and pre monsoon values of water quality parameters were taken for comparison. Three years (2009, 2010, and 2011) of data were taken to study the concentrations before the tunnel construction and it is shown in   Table 6.3. The data for the period 2012-2015 during tunnel construction phase and the data for the period 2016-2017 after the tunnel construction were also separated for both pre monsoon and post monsoon periods and are shown in   Table 6.4 and Table 6.5 respectively. The minimum and maximum values of water quality parameters before, during, and after the tunnel construction are shown in Figure 6.1, Figure 6.2, and Figure 6.3 respectively.

Comparing the water quality parameter pH during all the three phases on both the seasons it does not show any significant changes. All the values falls in the range of 7.5 -7.9. Total dissolved solids seemed to be slightly increased from before to after the construction. Before the construction, post monsoon and pre monsoon carried the value of 890 mg/l and 933 mg/l but after the construction the values appeared as 945 mg/l and 938 mg/l. During the construction, the presence of organic and inorganic substances tilted up in the excavation, because of the Total Dissolved solids seemed to be very high with the range of 1077 mg/l to 1069 mg/l during post monsoon and pre monsoon periods.

Total hardness value did not seem to have changed much in all the three phases but after the construction the values had gone to the range of 776 mg/l in the post monsoon and 770 mg/l in the pre monsoon period. It might have occurred due to the addition of various minerals used in various stages of operation and their getting accumulated. An increased value of total hardness was noted. Similarly, chloride concentrations also showed the same trend, it had not changed much in the early two phases but after the construction it showed an increased value of 874 mg/l and 856 mg/l in the post monsoon and pre monsoon periods. But, fluoride value showed a reduced value of about 0.47 and 0.42 in the post monsoon and premonsoon seasons. Earlier in the before and during the stages of construction, its values seemed to be 0.5 to 0.54.Fluroide is an earth mineral that occurs naturally in the surface. Decreased value of fluoride may be due to the breaking up of those fluoride minerals present below the surface of the soil.

Table 6.3 Minimum and Maximum values of water quality parameters in ground water samples before tunnel construction

Parameters	Pre-monsoon samples(mg/l)			Post-monsoon samples(mg/l)		
	Min	Max	Mean	Min	Max	Mean
pH	7.2	8.4	7.8	7.0	8.2	7.5
TDS	290	2481	933	287	2424	890
TH	175	560	349	101	660	349
Cl	98	735	361	96	678	307
Fl	0.18	0.86	0.51	0.15	0.82	0.44

Table 6.4   **Minimum and Maximum values of water quality parameters in ground water samples during tunnel construction**

Parameters	Pre-monsoon samples(mg/l)			Post-monsoon samples(mg/l)		
	Min	Max	Mean	Min	Max	Mean
pH	6.9	8.5	7.7	6.5	8.3	7.6
TDS	465	2019.5	1069	450	2117	1077
TH	150	625	339	198	625	333
Cl	121	780	308	101	734	306
Fl	0.30	1.02	0.54	0.18	0.92	0.56

Table 6.5   **Minimum and Maximum values of water quality parameters in ground water samples after tunnel construction**

Parameters	Pre-monsoon samples(mg/l)			Post-monsoon samples(mg/l)		
	Min	Max	Mean	Min	Max	Mean
pH	7.0	9.3	7.8	7.1	9.4	7.9
TDS	345	1800	938	350	1800	945
TH	348	1290	770	348	1292	776
Cl	258	1532	856	250	1789	874
Fl	0.13	0.85	0.42	0.13	0.86	0.47

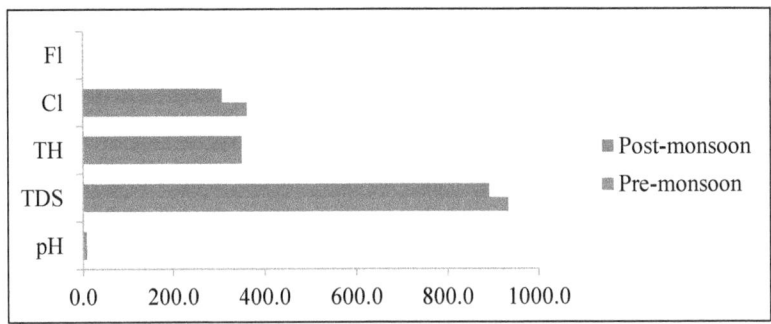

Figure 6.1 **Minimum and Maximum values of water quality parameters in ground water samples before tunnel construction**

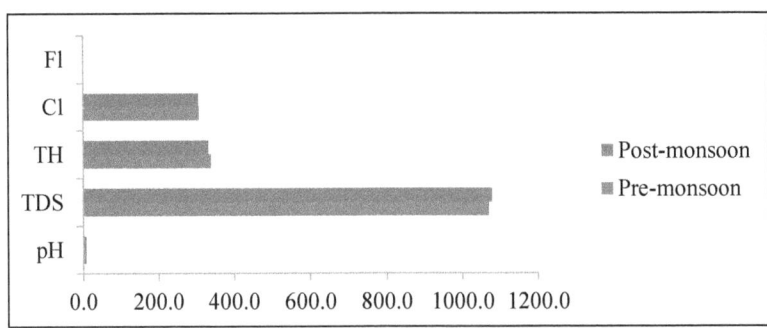

Figure 6.2 **Minimum and Maximum values of water quality parameters in ground water samples during tunnel construction**

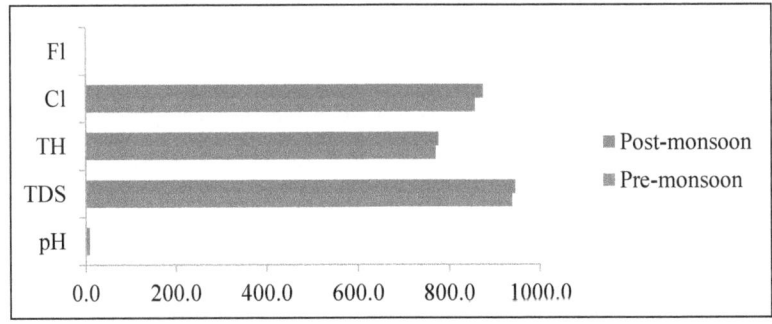

Figure 6.3 **Minimum and Maximum values of water quality parameters in ground water samples after tunnel construction**

6.3 SPATIAL ANALYSIS OF WATER QUALITY

The spatiotemporal behavior of the groundwater quality using GIS was carried in the study area. The spatial analyst module of Arc GIS 9.3 was used for this analysis. The data base was created in the Excel spread sheets for the premonsoon and post monsoon values of the periods before construction, during construction and after construction and the mean values was calculated.

Inverse Distance Weighted (IDW) interpolation technique was used in the analysis. IDW is an algorithm for spatially interpolating or estimating values between measurements. Each value estimated in an IDW interpolation is a weighted average of the surrounding sample points. Weights are computed by taking the inverse of the distance from an observation location to the location of the point being estimated.

In a comparison of several different deterministic interpolation procedures, using IDW with a squared distance term yielded results most consistent with original input data. This method is suitable for datasets where the maximum and minimum values in the interpolated surface commonly occur at sample points (ESRI 2002).

The spatial analysis for the water quality parameters such as pH, Total dissolved solids, Total Hardness, Chloride and Fluoride was done and compared. The spatial analysis was done for the mean pre monsoon and post monsoon values and it was compared for before(2009-2011), during(2012-2015), and after(2016-2017) the construction of metro rail corridors.

6.3.1   pH

Before the construction of metro rail corridor, the premonsoon value showed that the pH value around the two underground corridors was greater than 7.5 except in the corridor I; the stretch from Anna Nagar to Thirumangalam showed less than 7.5.During the construction, the pH values in the corridor I, the stretch that had pH value less than 7.5 now increased upto KMC from Thirumangalam while the rest of the status remained the same. But after the construction, the stretch from Anna Nagar to Thirumangalam beyond the extent carried the value less than 7.5. And also in the corridor II, the stretch from Teynampet to Saidapet carried the value less than 7.5.

Similarly, the post monsoon values before the construction, due to the recharge occurred by the rainfall in the wells, the pH concentration had dissolved,50% of the area from the mid of the study area towards to the West side of the study area took the value less than 7.5 and the remaining area had more than 7.5.During the construction, pH value disturbance that has occurred in the premonsoon period existed in the postmonsoon period too. After the construction, almost the value greater than 7.5 existed in the two underground corridors, only a small extent from Teynampet to Saidapet carried the value less than 7.5 in the corridor II and also small area towards North side of the corridor I carried the value less than 7.5.

It clearly shows that the water quality parameters are greatly affected by the tunneling process and also due to rainfall, the concentration level gets changed. The spatial analysis of water quality parameter pH before, during and after the construction during both premonsoon and post monsoon is shown in Figure 6.4.

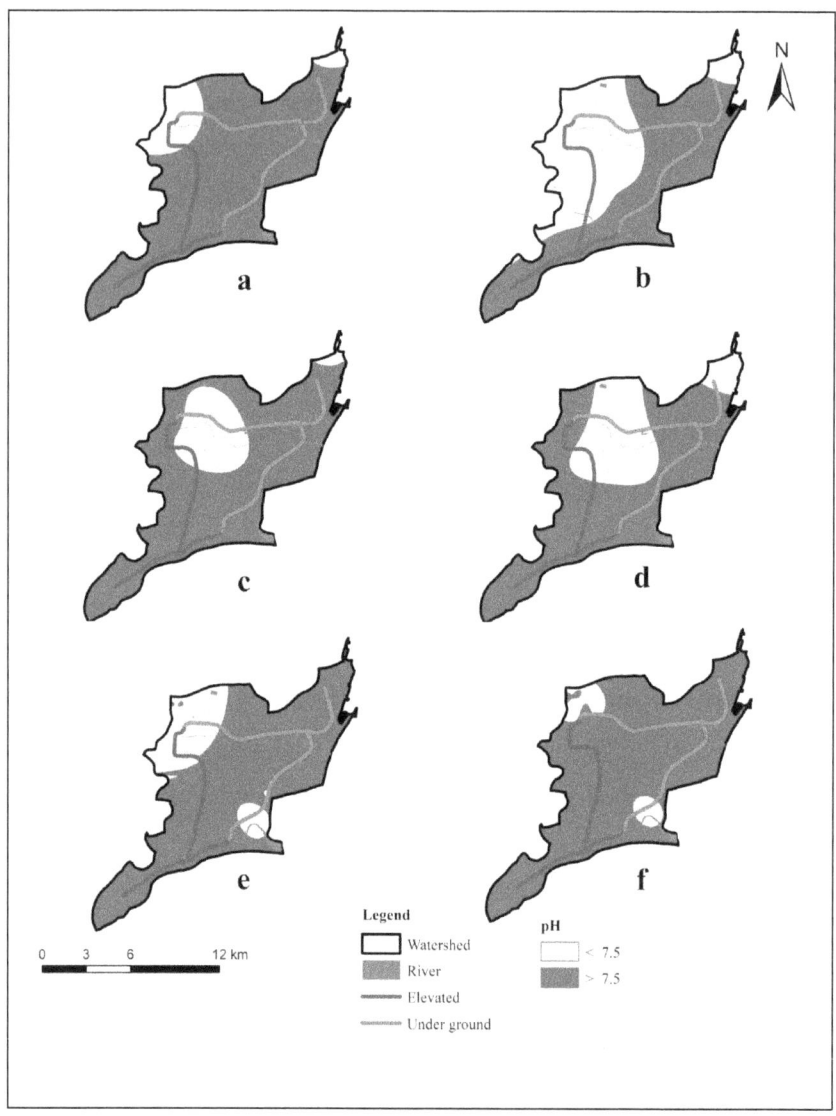

Figure 6.4   Spatial Analysis of pH (a) Before Construction, premonsoon
(2009-2011), (b) Before Construction, postmonsoon (2009-2011)
(c) During Construction, premonsoon, (2012-2015) (d) During
Construction , postmonsoon (2012-2015) (e) After Construction,
premonsoon, (2016-2017) (f)After Construction, postmonsoon,
(2016-2017)

6.3.2   Total Dissolved Solids

Total dissolved solids is meant for the presence of combined organic and inorganic content in colloidal or suspended form. Its permissible limit is from 500 to 1500 mg/l. Before the construction, the values less than 500 mg/l could be seen at the end of the Thirumangalm in corridor I and also the area below Saidapet in corridor II.The initial stretch, that is from Central to KMC in corridor I and Central to LIC in corridor II carried the higher range of greater than 1500 mg/l with the premonsoon values. During the construction the area with value less than 500 mg/l seemed to be not existing, even in the after phase construction. After the construction, the whole study area carried the value range from 500 -1500 mg/l except a small stretch from Teynampet to Saidapet that took more than 1500 mg/l.

The post monsoon values did not seem to be much varied from the premonsoon values in before, during, and after the construction phase.It could be due to the colloidal and suspended nature of both organic and inorganic substances. The colloidal tends to get blended with the soil and due to the sealants used in the tunneling process the value of the Total Dissolved solids gets altered.

It is clearly seen that once the nature of the soil below the ground surface gets disturbed through this trenching, the layer properties of the soil get modified and also the index properties of the soil get changed along with physical factors such as rainfall and temperature, and the concentration of the parameter varies. The spatial analysis of water quality parameter Total dissolved solids before, during, and after the construction during both premonsoon and post monsoon is shown in the Figure 6.5.

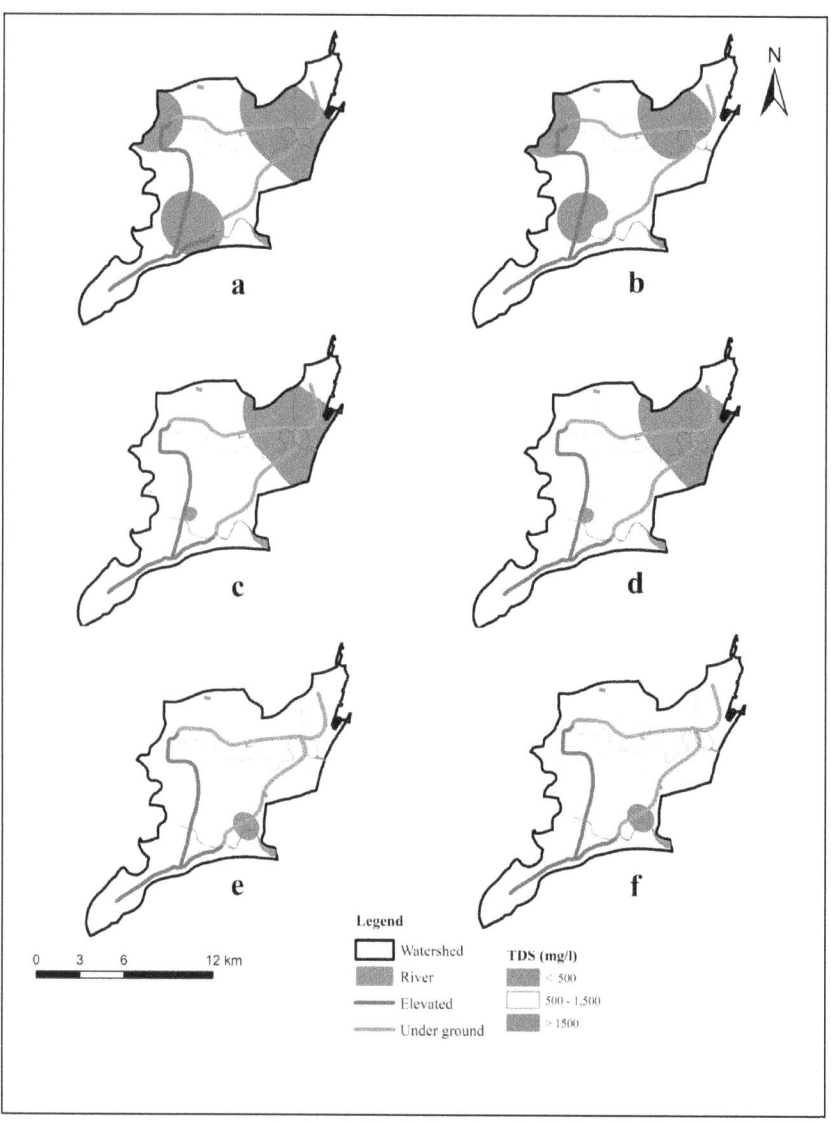

Figure 6.5 **Spatial Analysis of TDS (a) Before Construction, premonsoon (2009-2011), (b) Before Construction, postmonsoon (2009-2011) (c) During Construction, premonsoon, (2012-2015) (d) During Construction , postmonsoon (2012-2015) (e) After Construction, premonsoon, (2016-2017) (f)After Construction, postmonsoon, (2016-2017)**

6.3.3   Total Hardness

It is the presence of calcium and magnesium salts present in water. The water with high mineral content is said to be hard water and the value will be greater than 500 mg/l. If it is less than 200 mg/l it is said to be soft water. The pre monsoon values showed that before construction all the areas carried the value from 250- 500 mg/l except the stretch from Central to KMC that carried more than 500 mg/l in the corridor I , that is from Central to Thirumangalam. In corridor II, the stretch below Saidapet carried the range below 250 mg/l. During the construction, the stretch from Shenoi Nagar to the end of Thirumangalam became less than 250 mg/l. In corridor II, from Central to LIC the East side of the corridor got affected with higher values of more than 500 mg/l. But after the construction, may be due to the addition of chemicals it contained the composition of both calcium and magnesium and the whole study area found the Total Hardness to be more than 500 mg/l.

The post monsoon values also revealed the same status that after the construction the study area got affected with higher values. During the construction the initial stretch in both the underground corridors, that is from Central to KMC in corridor I and Central to LIC in corridor II got affected with higher Total Hardness values. From KMC to Thirumangalam the lower hardness values existed. The same scenario existed even in 'before the construction' of post monsoon values.

The reason behind the higher values of Total hardness 'after the construction' phase in both pre monsoon and post monsoon could be only the use of cementious materials that contain the composition of both calcium and magnesium minerals. The minerals which have got accumulated during the construction of the corridors have yielded the higher values of Total hardness in the study area. The spatial analysis of water quality parameter TH before, during, and after the construction during both premonsoon and post monsoon is shown in Figure 6.6.

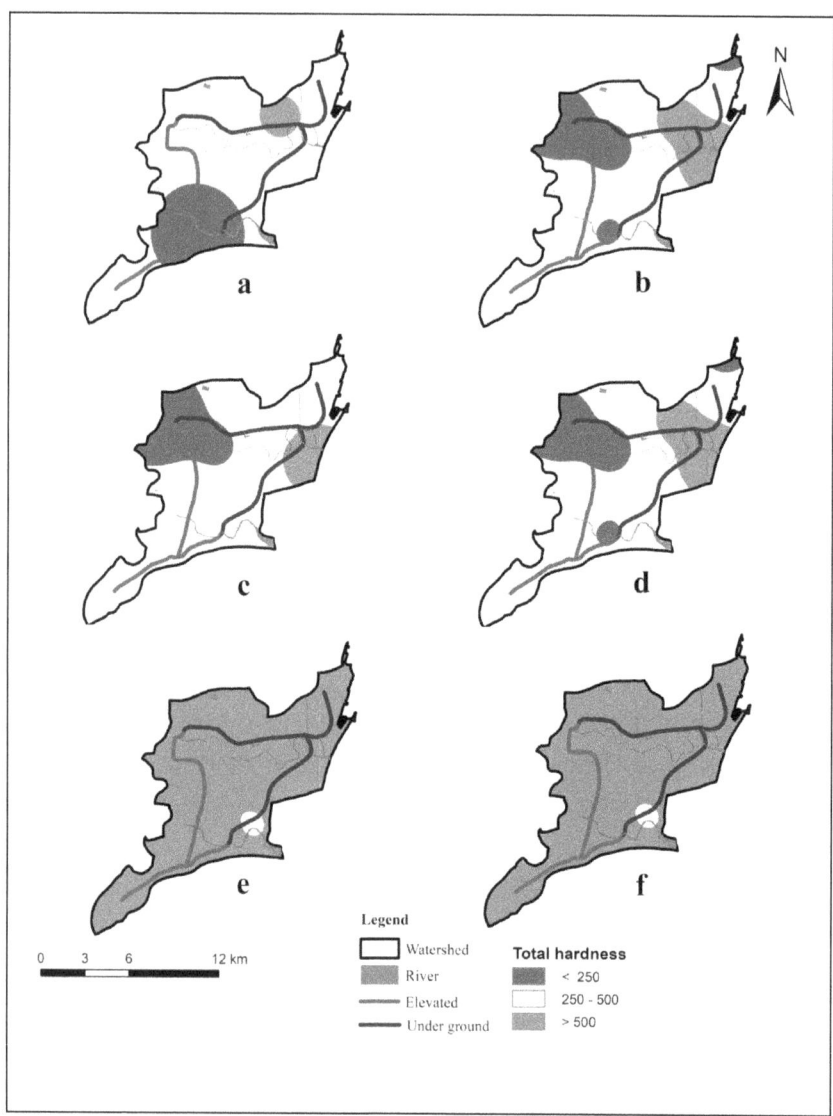

Figure 6.6   Spatial Analysis of TH (a) Before Construction, premonsoon
(2009-2011), (b) Before Construction, postmonsoon (2009-2011)
(c) During Construction, premonsoon, (2012-2015) (d) During
Construction , postmonsoon (2012-2015) (e) After Construction,
premonsoon, (2016-2017) (f)After Construction, postmonsoon,
(2016-2017)

6.3.4   Chloride

The pre monsoon and the post monsoon values during and after the construction showed the same effect on both sides of the corridors.Initial stretch in both the corridors, i.e., from Central to KMC in corridor I and Central to LIC in corridor, showed the higher range of concentrations that is greater than 600 mg/l. The areas which did not get affected are stretches that include from Shenoi Nagar to Thirumangalam in corridor I and the area below Saidapet in the corridor II. The rest of the areas showed the range of 200 -600 mg/l in both the pre monsoon and postmonsoon seasons.

After the construction, the whole study area carried the concentration more than 600 mg/l in both the seasons except the stretch on either side of the KMC in the corridor I that carries the value of 200- 600 mg/l. But before the construction the premonsoon value showed a small stretch down the corridor II with low values and initial stretch in both the corridors got affected with higher concentrations.

Sodium chloride is the main source of chloride. The water source that was used during the tunneling process and also some of the chemical compounds that were used during the construction could be the source of the chloride and became the reason for showing increased concentration after the construction. The spatial analysis of water quality parameter Chloride before, during, and after the construction during both premonsoon and post monsoon is shown in the Figure 6.7.

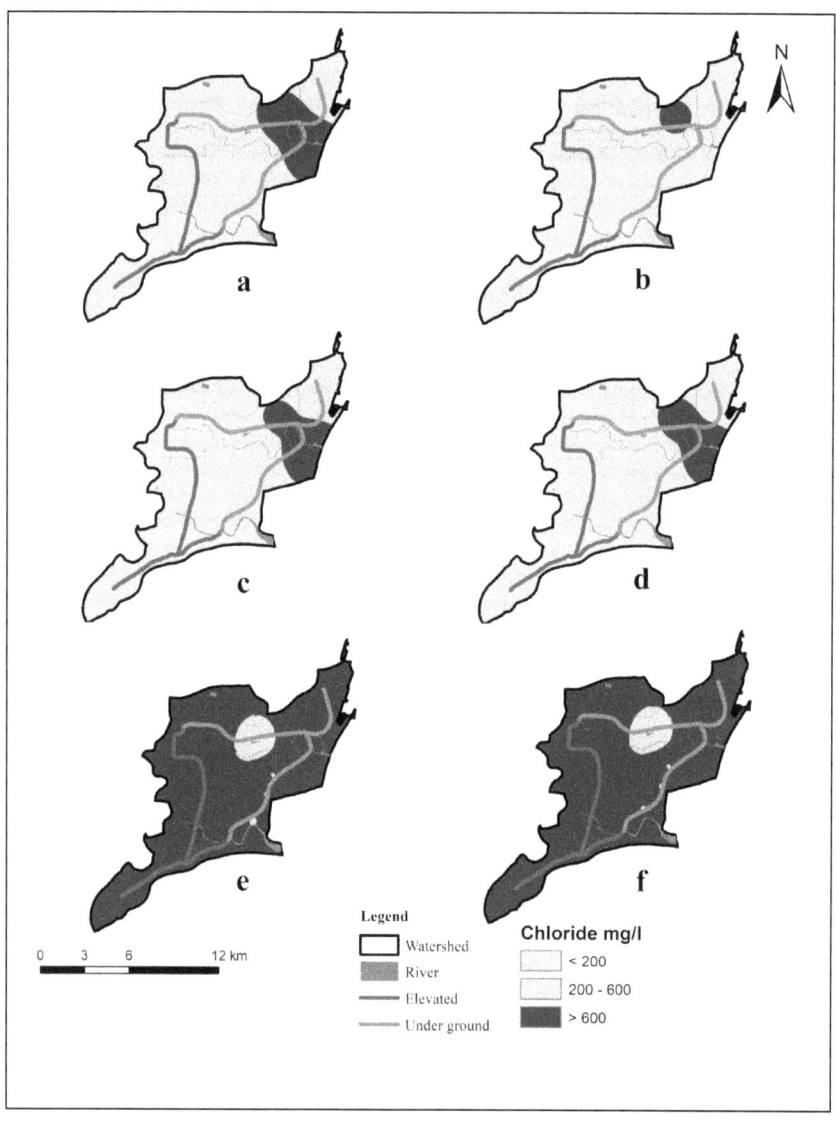

Figure 6.7   Spatial Analysis of Cl (a) **Before Construction, premonsoon (2009-2011), (b) Before Construction, postmonsoon (2009-2011) (c) During Construction, premonsoon, (2012-2015) (d) During Construction , postmonsoon (2012-2015) (e) After Construction, premonsoon, (2016-2017) (f)After Construction, postmonsoon, (2016-2017)**

6.3.5   Fluoride

Fluoride is found in nature abundantly in the soil. The fluoride salts become higher in quantity when the fluoride minerals are present in the earth's crust. The excess of fluoride present in the soil, when consumed in excess, results in fluorosis. Both the underground corridors carried the range of fluoride from 0.4 to 0.6 mg/l. Before the construction the pre monsoon values less than 0.4 mg/l existed in both the underground corridors except the initial stretch of the corridor II showing from 0.4- 0.6 mg/l. During the construction, the corridor II shows some areas less than 0.4 mg/l. After the construction, the area from Tondiarpet to the Central carried the value less than 0.4 mg/l and the rest carried the range of 0.4 to 0.6 mg/l.

The post monsoon values showed that the value less than 0.4 mg/l always existed in the same location as that of the pre monsoon values. Only during the construction the higher range that was more than 0.6 mg/l existed at the end of Thirumangalam in corridor I and down to the Saidapet area in corridor II. After the construction the higher concentrations vanished and only the range 0.4 to 0.6 mg/l almost existed in both corridors.

Water on either side of the KMC in corridor I seemed to be affected with high fluoride value. Trenching technology involved in the breaking of the soil structure, could be the reason showing higher concentrations during the construction phase. Even it shows the higher concentration of more than 0.8, it is still under the permissible limit on both underground corridors. The spatial analysis of water quality parameter (Fluoride) before, during, and after the construction during both premonsoon and post monsoon is shown in Figure 6.8.

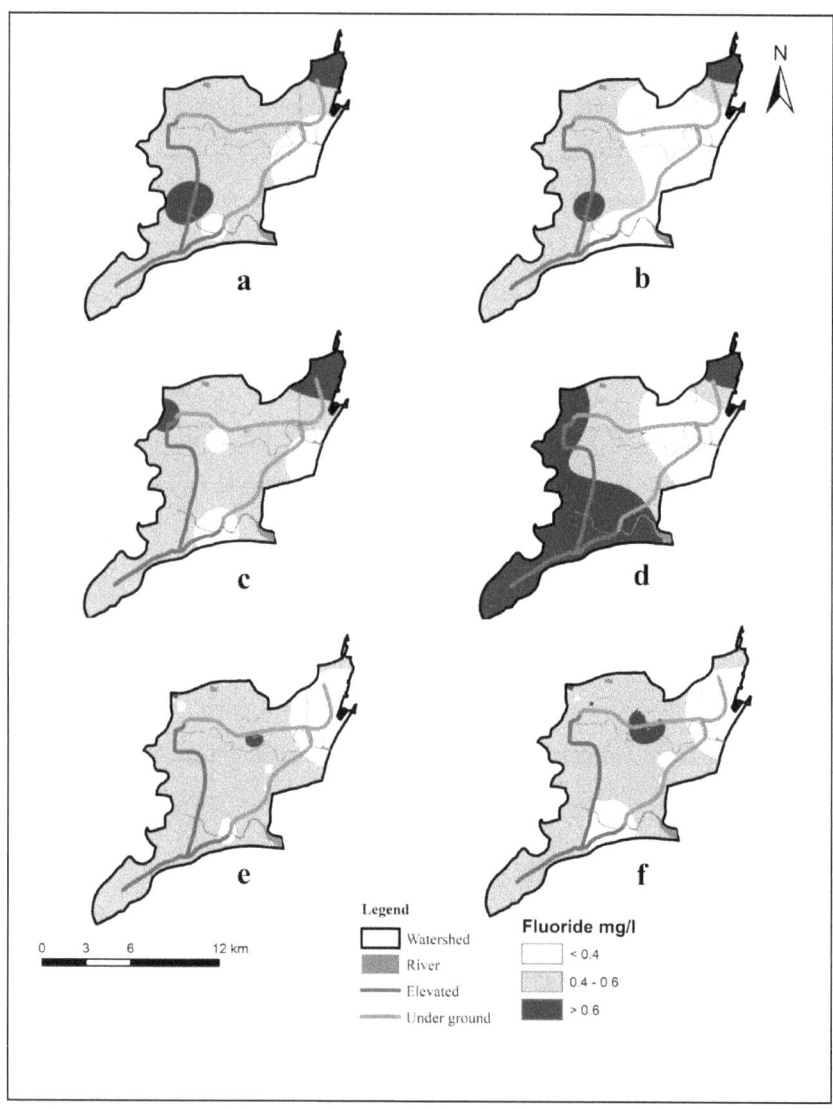

Figure 6.8   Spatial Analysis of Fl (a) Before Construction, premonsoon
(2009-2011), (b) Before Construction, postmonsoon (2009-2011)
(c) During Construction, premonsoon, (2012-2015) (d) During
Construction , postmonsoon (2012-2015) (e) After Construction,
premonsoon, (2016-2017) (f)After Construction, postmonsoon,
(2016-2017)

6.4   WATER QUALITY INDEX DETERMINATION

Water quality index is one of the most effective tools to communicate information on the quality of any water body. This technique uses a mathematical equation to transform a large number of water quality data into a single number. It integrates the complex data and generates a single figure to find out the water quality status of the ground water.

As described in the methodology the water quality parameters of the two phases of the construction are separated and the parameters are calculated for the water quality status as per the water quality index determination procedure and it is shown in Table 6.6 and Table 6.7. It clearly shows that water quality after the construction has deteriorated.

All the parameters show the minimum value and the maximum value which exceed the limit of the standards. Water quality index methodology adopted to the water quality parameters reveals that some of the parameters show a major concentration difference like Total dissolved solids and Total alkalinity but overall the concentrations are not much changed. The Water Quality Index Value for the period 1995-2008 has been calculated as 52.63 and for the period 2009-2017 it has been calculated as 54.321. These values clearly show that the mobilization of concentrations and the water quality status have not been same but slowly getting deteriorated due to the tunneling effect.

Table 6.6 Water Quality Index Determination for 2009 -2017

S.No	Parameters	Indian Standards	Mean Value	Min Value	Max Value	Weightage (w_i)	Relative weight (W_i)	Quantity Rating (q_i)	Sub Index (SI_i)
1	pH	6.5 -8.5	7.7	7.4	8.2	4	0.1212	90.58	10.978
2	EC,µs/cm	500-2000	1800.5	943.3	3952.3	4	0.1212	90.025	10.911
3	TDS,ppm	500-2000	1345.9	753	2900	4	0.1212	67.29	8.155
4	TH,mg/l	300-600	342.7	200	662.5	3	0.0909	57.12	5.192
5	Ca^{2-},mg/l	75-200	61.1	29	123.4	2	0.0606	30.55	1.851
6	Mg^{2-},mg/l	30-100	46.9	24	124.1	2	0.0606	46.9	2.842
7	TA, mg/l	200-600	120	75	225	3	0.0909	21.5	1.954
8	Cl^{-}, mg/l	250-1000	326.7	110	555.1	3	0.0909	32.67	2.969
9	F,mg/l	1-1.5	0.6	0.3	0.72	4	0.1212	40	4.848
10	So_4^{2-},mg/l	200-400	152.5	43	680.7	4	0.1212	38.125	4.621
						$\Sigma w_i = 33$			ΣSI_i =54.321

Table 6.7 Water Quality Index Determination for 1995-2008

S.No	Parameters	Indian Standards	Mean Value	Min Value	Max Value	Weightage (w_i)	Relative weight (W_i)	Quantity Rating (q_i)	Sub Index
1	pH	6.5 - 8.5	8	7.5	8.4	4	0.1212	94.11	11.406
2	EC ,μs/cm	500-2000	1770.6	900.3	4167	4	0.1212	88.53	10.729
3	TDS,ppm	500-2000	932.4	564.3	2327.2	4	0.1212	46.62	5.65
4	TH,mg/l	300-600	375.6	243.9	722	3	0.0909	62.6	5.69
5	Ca^{2+},mg/l	75-200	68.1	31	123.4	2	0.0606	34.05	3.036
6	Mg^{2-},mg/l	30-100	50.1	32.6	128.9	2	0.0606	50.1	3.036
7	TA, mg/l	200-600	187	85	300	3	0.0909	31.17	2.833
8	Cl^{-}, mg/l	250-1000	338.3	100.3	823.7	3	0.0909	33.83	3.075
9	F,mg/l	1-1.5	0.5	0.3	0.8	4	0.1212	33.33	4.039
10	So_4^{2-},mg/l	200-400	135.5	86.6	401.4	4	0.1212	33.875	4.106
						$\Sigma w_i = 33$			ΣSI_i =52.63

6.5   **QUALITY ANALYSIS OF PRIMARY WELLS**

The datas which were collected from the 20 primary observations wells were analyzed for their variations in the two underground developments. The five parameters, namely Chloride , Fluoride, pH, Total Dissolved Solids and Total Hardness were compared with respect to either side of the underground barrier constructed below the surface of the soil.

6.5.1   **Comparison of Water Quality Parameters in Opposite Wells**

In the corridor I that is from Central to Thirumangalam during the post-monsoon period pH value carried a minimum of 7.1and maximum of 8.4 in the North side of the corridor, whereas a minimum of 7.1and maximum of 9.4 existed in the South side of the corridor. Again in the premonsoon period, the South side of the corridor carried the higher value of pH upto 9.3 whereas in the North side the maximum value was only 8.2. Hence in both the seasons South side of the corridor got affected with higher pH value. Similarly, in the corridor II that is from Washermenpet to Saidapet East side of the corridor got affected with the higher value of 8.9 on both seasons. West side of the corridor showed the maximum of 8.5 in the post monsoon and maximum of 7.1 during the premonsoon period.

In the post monsoon, water quality parameter TDS presence in the corridor I was found to be higher in the South side of the corridor. Its minimum to maximum range was from 689 mg/l to 1300 mg/l. In the North side it was only from 470 mg/l to 850 mg/l. In the post monsoon, the values in the South side seemed to be with the higher range 690 mg/l and 1305 mg/l and in the North side it showed the lower range of about 469 mg/l and 853 mg/l. In both sides the TDS values were within the range of 500 -1500 mg/l as per WHO standards. Similarly, for corridor II, East side of the corridor showed a maximum value upto 1800 mg/l during both seasons which was higher than the prescribed limits of WHO.

The total hardness values were found to be higher and exceeding the limits of WHO, which is about 75-300 mg/l. Comparatively, South side of the corridor I on both seasons showed higher values from 650-1290 mg/l. Similarly, in the corridor II , West side of the corridor showed higher value ranges about 350-1089 mg/l during both seasons. For the corridor I ,Central to Thirumangalam, the North side of the corridor got affected with the higher chloride value that ranged about 400-1782 mg/l which is much higher than the prescribed limit of WHO range of 200-600 mg/l. South side of the corridor too showed a higher range but it existed upto 1200 mg/l. In corridor II, the West side of the corridor showed higher values when compared with the East side of the corridor. On both side it exceeded the prescribed limit. West side showed the maximum of1388 mg/l and East side showed the maximum of 1026 mg/l in both the seasons.

Water quality parameter (Fluoride) found to be higher, with the value ranging from 0.34 to 0.84 and 0.29 to 0.85 on the North side of the corridor during postmonsoon and premonsoon periods. Similarly the West side of the corridor carried the maximum values 0.32 to 0.56 in the postmonsoon and 0.3 to 0.48 in the premonsoon periods. In either side of the corridor the fluoride was found to be within the prescribed limit of 0.6 to 1.5 WHO standards.

Water quality parameters pH, Total dissolved solids and Total hardness showed higher values on the south side of the corridor. It could be due to the decreased water level pattern on the south side of the corridor. Higher water levels may dissolve the concentrations and shows lower values. Fluoride was found to be higher on the North side of the corridor I and West side of the corridor II. It could due to the geological formation existed on the

North side and the West side of corridor I and corridor II. Again it could have been influenced by the artificial weathering that was breaking out of the rocks during the tunneling process. Chloride concentrations are generally not absorbed by the soil and it is only chloride ions that are found in water. Hence the flow direction favours the contaminant transport on the respective side of the corridors. The comparisons of East and West side of corridor II and North and south side of the corridor I are shown in Figure 6.9 and Figure 6.10 respectively.

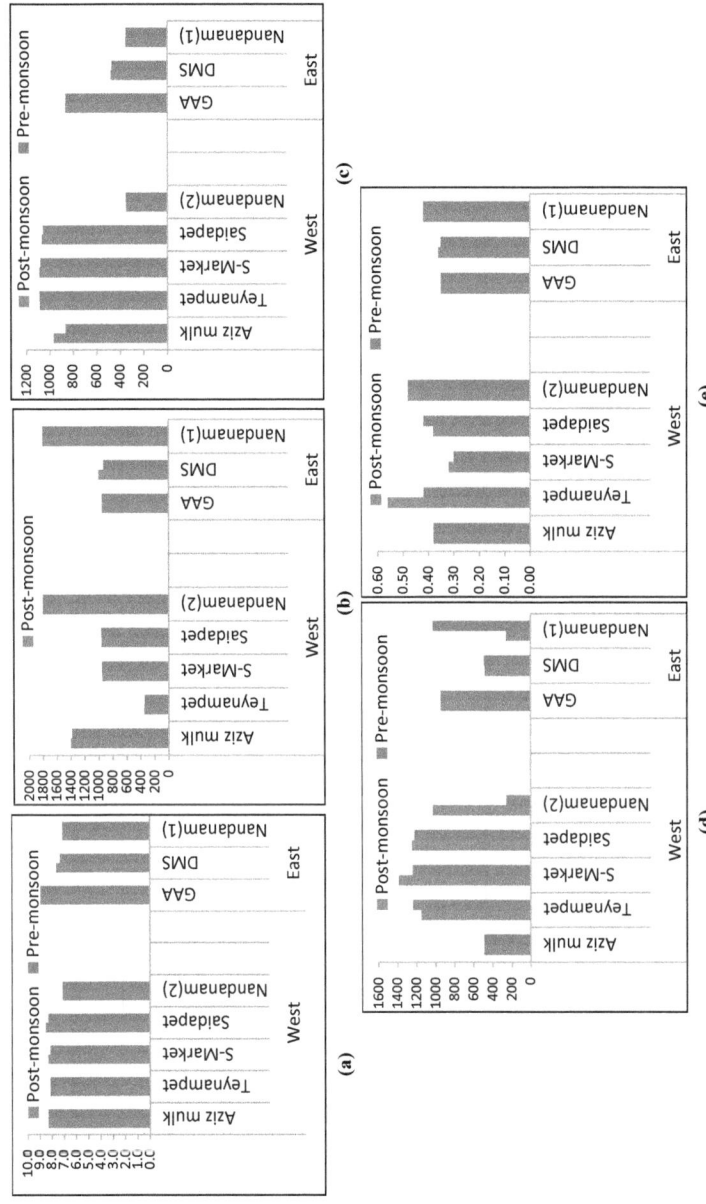

Figure 6.9 Comparison of West and East side, Pre-monsoon and Post-monsoon water quality parameters(mg/l) of corridor, Washermenpet to Saidapet after tunnel construction (a)pH(b)TDS(c)TH(d)Cl(e)Fl

164

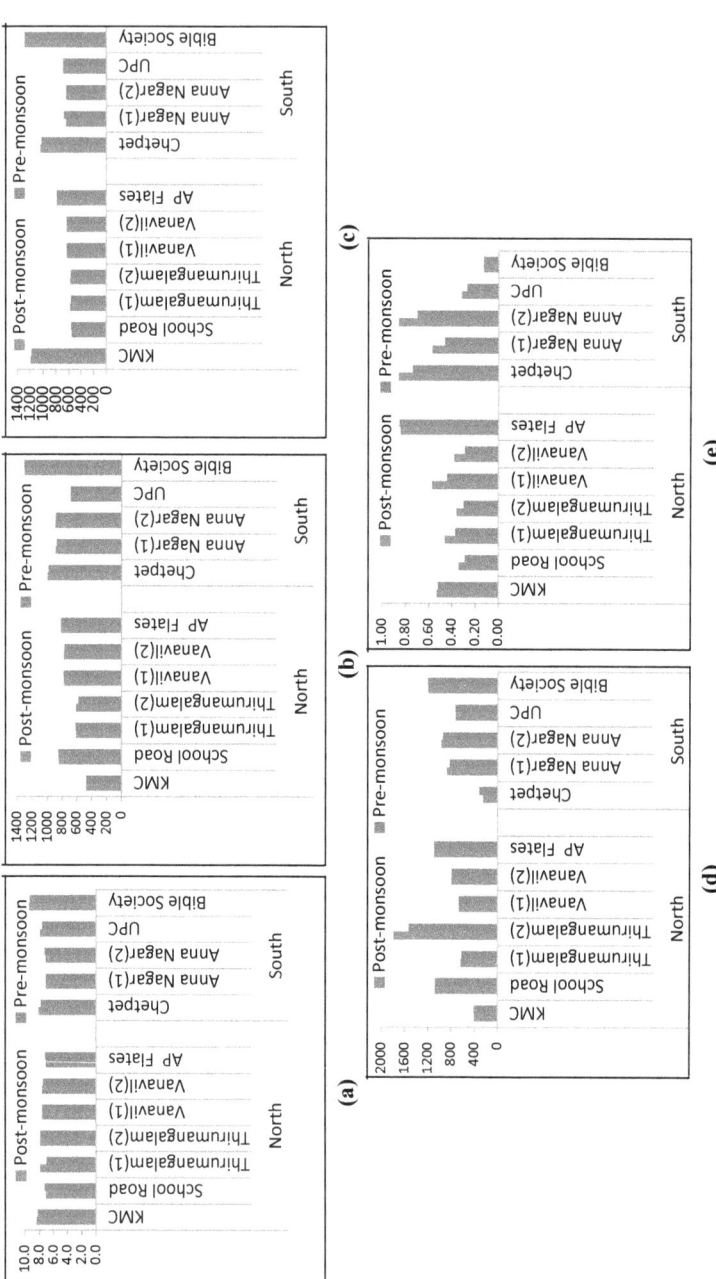

Figure 6.10 Comparison of North and South side, Pre-monsoon and Post-monsoon water quality parameters (mg/l) of corridor, Central to Thirumangalam after tunnel construction (a)pH(b)TDS(c)TH(d)Cl(e)Fl

6.6 GROUND WATER QUALITY MODELING

Ground water quality modeling was done with the parameters like Calcium, Chloride, Electrical Conductivity, Fluoride, pH, Sulphate, Total Dissolved Solids and Total Hardness. The concentration levels of the above parameters were given as inputs along with the hydro geological system given for the ground water flow modeling. The change in the concentration levels due to tunneling was predicted in the modeling.

6.7 PROCEDURE FOR CONTAMINANT TRANSPORT MODELING

The model inputs and the conceptualization of the Water Flow Model were the same as that for the contaminant transport modeling. The eight parameters which are likely to affect the Water quality were given as input for the concentration and the contaminant transport was calibrated, validated, and predicted for the same phases of running the model in the water flow model. In each phase, the progress of tunneling was executed in the base map as wall constructed and the movement of the contaminant was predicted.

6.7.1 Model Calibration

The model was run and calibrated for transient state condition. The calibration was carried for the time period from 1995 to 2005 (3650 days). The dispersivity and the recharge concentration were varied so as to obtain reasonably good correlation between calculated and observed concentration. From the graph of calculated vs. observed concentration, it can be seen that the calculated concentration matches well the observed concentration at 99 % confidence level. The calculated and the observed concentration for a period of 10 years and some of the observation wells given in Figures 6.11 (a), (b), (c), (d), (e), and (f) show fairly good agreement.

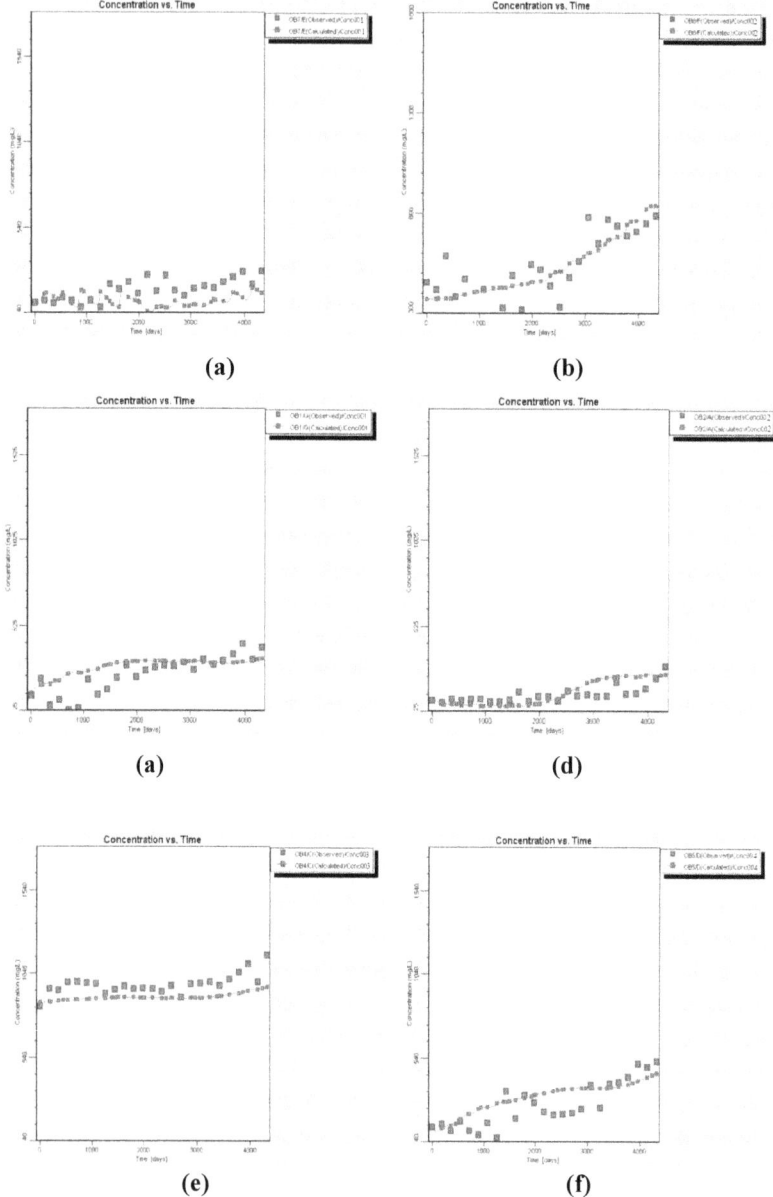

Figure 6.11 Calibrated   concentrations   for   observation   wells
(a) Vepery (b) Chepauk (c) Saidapet (d) Aminjikarai
(e) Thirumangalam (f) K.K. Nagar

6.7.1.1   Chloride

The model was run and calibrated for transient state condition. The calibration was carried for the time period from 1995 to 2005 (3650 days) and is shown in Figures 6.12 (a) 730 days and (b) 2190 days. The dispersivity and recharge concentration were varied so as to obtain reasonably good correlation between the calculated and the observed concentrations. From the graph of calculated vs observed concentration, it can be seen that the calculated concentration matches well the observed concentration at 95 % confidence level.

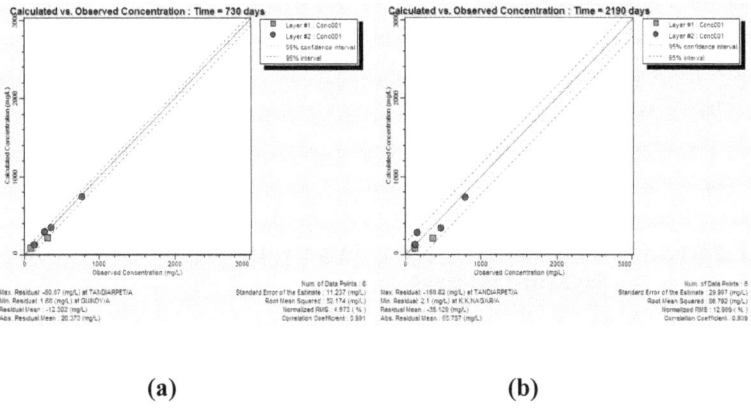

(a)   (b)

Figure 6.12   Calculated   vs   Observed   Concentration   for   Chloride
(a) 730 days (b) 2190 days

6.7.1.2   Fluroide

The model was run and calibrated for transient state condition. The calibration was carried for the time period from 1995 to 2005 (3650 days) and is shown in Figure 6.13 (a) and (b) for 1095 days and 3521 days. The Fluoride parameter with the concentration ranged from 0.1 mg/l to 0.8 mg/l . The dispersivity and the recharge concentration were varied so as to obtain reasonably good correlation between the calculated and the observed concentration. From the graph of calculated vs. observed concentration, it can be seen that the calculated concentration matches well the observed concentration at 95 % confidence level.

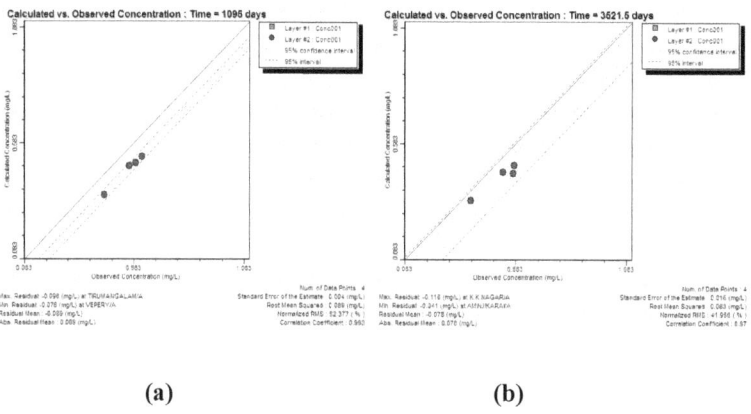

(a)   (b)

Figure 6.13   Calculated vs Observed Concentration for Fluroide (a) 1095 days (b) 3521 days

6.8   MODEL VALIDATION

6.8.1   Total Hardness

Following calibration, the contaminant transport model was validated for a period of 3 years from 2006 to 2009.The total hardness of concentration 100 mg/l to 800 mg/l was observed in the validation.   The observed concentration showed a good correlation with the calculated concentration for all the observation wells for the days of 4080 days and 5110 days as shown in Figures 6.14 (a) and (b) at 95% confidence interval.

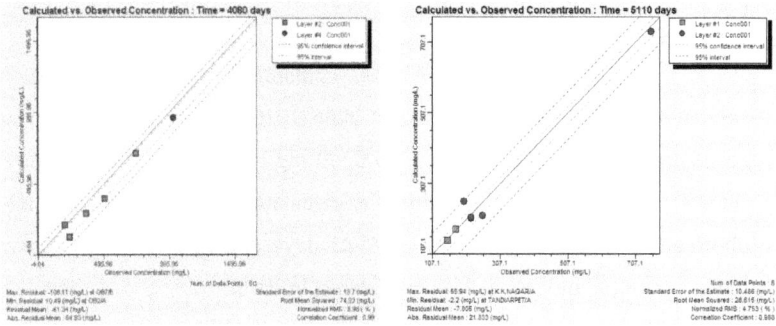

**Figure 6.14 Calculated vs Observed Concentration for Total Hardness
(a) 4080 days (b) 5110 days**

6.8.2   pH

Following calibration, the contaminant transport model was validated for a period of 3 years from 2006 to 2009.The Contour map of the pH of concentration 7.30 to 8.2 is shown in Figure 5.57.   The observed concentration shows a good correlation with the calculated concentration for all the observation wells for the days of 4745 days and 5110 days as shown in Figures 6.15 (a) and (b) at 95% confidence interval

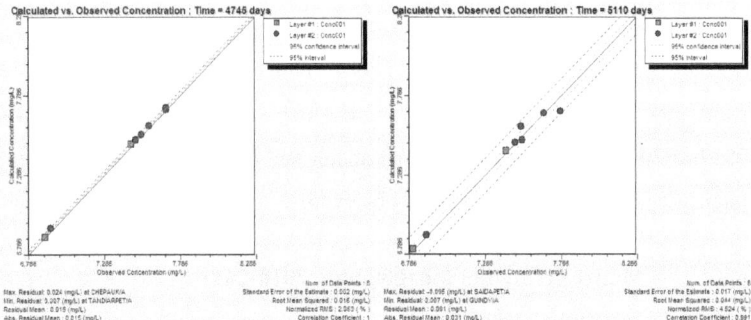

**Figure 6.15   Calculated vs Observed Concentration for pH (a) 4745 days
(b) 5110 days**

6.8.3   Total Dissolved Solids

Following calibration, the contaminant transport model was validated for a period of 3 years from 2006 to 2009.The total dissolved solids concentration ranged from 0 mg/l to 2500 mg/l .The observed concentration showed a good correlation with the calculated concentration for all the observation wells for the days of 4380 days and 5110 days as shown in Figures 6.16 (a) and (b) at 95% confidence interval.

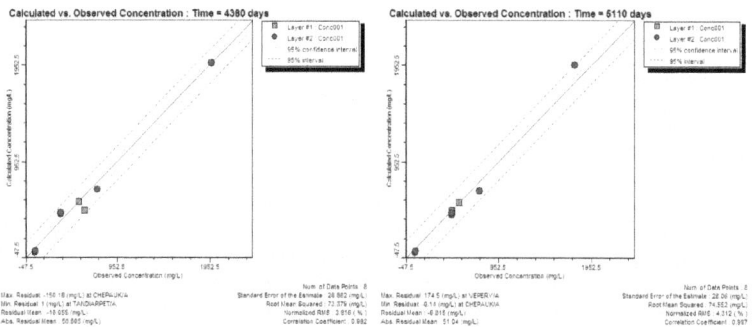

Figure 6.16   Calculated vs Observed Concentration for Total Dissolved Solids (a) 4380 days   (b) 5110 days

6.9   MODEL PREDICTION

Model prediction was done under various phases for each parameter such as 2010 -2011, 2012-2013, 2014-2015, 2015-2016 & 2017-2020. The behavior of each parameter and its concentration changed and intensities for various phases were observed.

6.9.1   Model Prediction for the phase 2010-2011

6.9.1.1   Chloride

The corridor stretch from Washermenpet to Saiḍapet carried the concentration in the range from 0 mg/l to 282.571 mg/l. To the end of the

Thirumangalam the concentration ranged from the minimum of 685.74 to 800 mg/l. From Anna Nagar tower to the Central the concentration ranged from 342.857 mg/l to 571.429 mg/l

The values have been predicted when the calculated and the observed concentrations showed the correlation coefficient of 0.964 at 95 % of confidence interval. The calculated and the observed concentrations of the chloride parameter for 5840 days are shown in Figure 6.17.

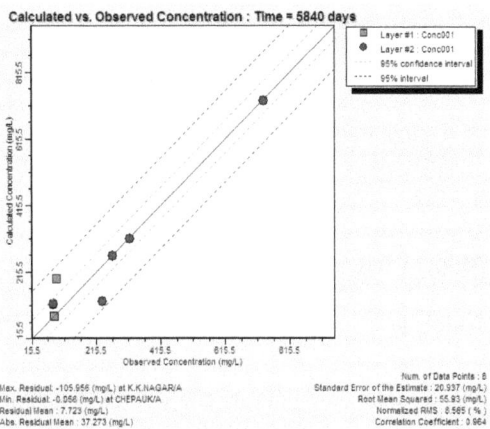

Figure 6.17   Calculated vs Observed Concentration for Chloride for 5840 days

6.9.1.2   Fluoride

In the corridor II, stretch from Washermenpet to Central carried the higher   concentration ranged from 0.729-08.00. Followed from Chepauk to DMS value ranged from 0.514 to 0.657. The corridor I fully came under the range of 0.514 to 0.657.

The values were predicted when the calculated and the observed concentrations showed the correlation coefficient of 0.868 at 95 % of confidence interval. The calculated and the observed concentrations of the chloride parameter for 5840 days are shown in Figure 6.18.

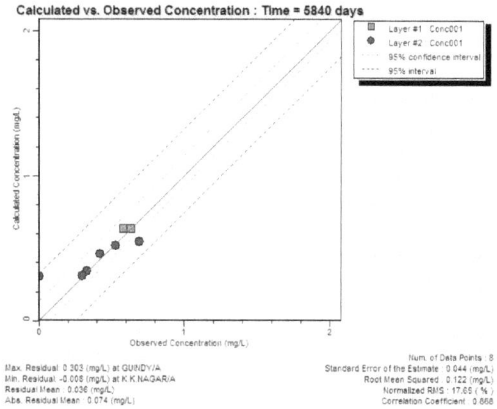

Calculated vs. Observed Concentration : Time = 5840 days

Figure 6.18 Calculated vs Observed Concentration for Fluoride for 5840 days

6.9.2   Model Prediction for the Phase 2012-2013

6.9.2.1   Chloride

The parameter of chloride is given for the phase of 2012-2013. The corridor stretch from Washermenpet to Saidapet carried the concentration in the range from 0 mg/l to 342.65 mg/l .For half of the stretch from Central to Thirumangalam, the concentrations were in the range from 457.35 mg/l to 685.7 mg/l and at the end of the stretch it increased to the concentration upto 800 mg/l. The Contour lines are closed enough to show the concentration values.

The values were predicted when the calculated and the observed concentrations showed the correlation coefficient of 0.997 at 95 % of confidence interval. The calculated and the observed concentrations of the chloride parameter for 6570 days are shown in Figure 6.19.

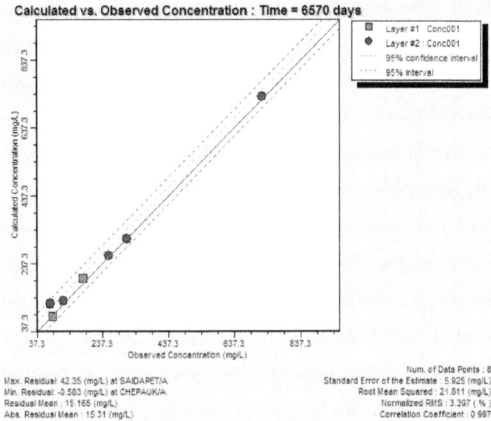

Figure 6.19 Calculated vs Observed Concentration for Chloride for 6570 days

6.9.2.2 Total Hardness

In corridor I ,the stretch from Thirumangalam to Aminjikarai, the values ranged from 400 -600 mg/l. Towards Central the concentration decreased from 300 mg/l to 100 mg/l. The Corridor II fully fell under the range of 100 mg/l to 300 mg/l. The values were predicted when the calculated and the observed concentrations showed the correlation coefficient of 0.999 at 95 % of confidence interval. The calculated and the observed concentrations of the chloride parameter for 6570 days are shown in Figure 6.20.

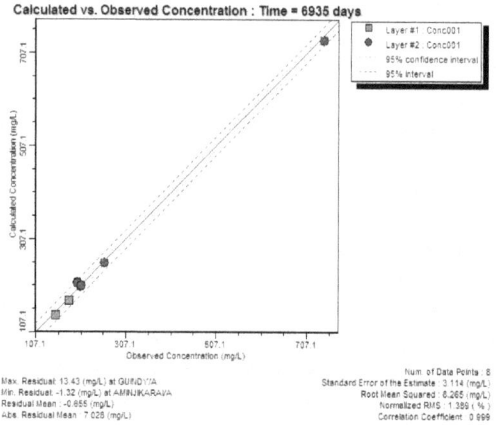

Figure 6.20 Calculated vs Observed Concentration for Total Hardness for 6935 days

6.9.3 Model Prediction for the Phase 2014-2015

6.9.3.1 Fluoride

The parameter of Fluoride is given for the phase of 2014-2015. The half of the corridor stretch from Saidapet to DMS carried the concentration in the range from 0.3 mg/l to 0.514 mg/l. From DMS to Washermenpet the concentrations were in the range from 0.586 mg/l to 0.729 mg/l and at the end of the stretch it increased to the concentration upto 0.8 mg/l. The corridor from Central to Thirumangalam had the concentrations within the range of 0.514 mg/l to 0.657 mg/l. The Contour lines are close enough to show the concentration values. The values were predicted when the calculated and the observed concentrations showed the correlation coefficients of 0.994 at 95 % of confidence interval. The calculated and the observed concentrations of the Fluoride parameter for 7300 days are shown in Figure 6.21.

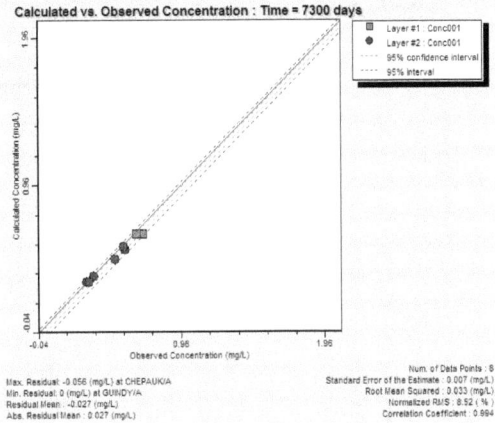

Figure 6.21 Calculated vs Observed Concentration for Fluoride for 7300 days

6.9.3.2 pH

The parameter of pH is given for the phase of 2014-2015.The stretch from Central to Vepery carried the concentration in the range from 7.943 to 7.547 . From Vepery to Tirumangalam it carries the smaller concentrations in the range from 7.429 to 7.300. The corridor from Central to Saidapet showed some higher concentrations in the range of 7.943 to 8.071. The Contour lines are very close concentration values in the corridor stretch Central to Tirumangalam. The values were predicted when the calculated and the observed concentrations shows the correlation coefficients of 0.998 at 95 % of confidence interval. The calculated and the observed concentrations of the pH parameter for 7300 days are shown in Figure 6.22.

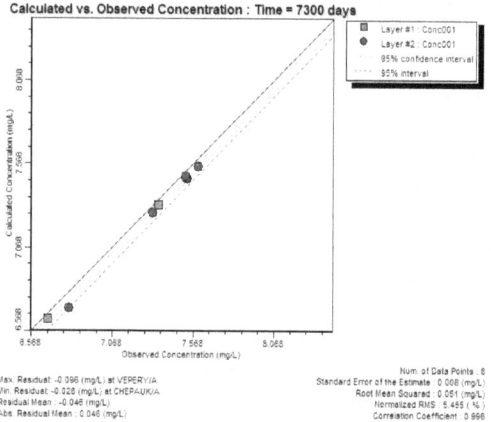

Figure 6.22 Calculated vs Observed Concentration for pH for 7300 days

6.9.4   Model Prediction for the Phase 2015-2016

6.9.4.1   Total dissolved solids

The prediction of the Total Dissolved solids parameter for the phase of 2015-2016 was made.The Centre part of the corridor from Central to Tirumangalam showed higher concentrations from 1071.429 mg/l to 1428.571 mg/l . The rest of the areas in both the corridors were found to be in the safer range of 0 to 714 mg/l. The Centre part of the corridor I showed a wider range lines. The values were predicted when the calculated and the observed concentrations showed the correlation coefficients of 0.984 at 95 % of confidence interval. The calculated and the observed concentrations of the Total dissolved solids parameter for 8210 days are shown in Figure 6.23.

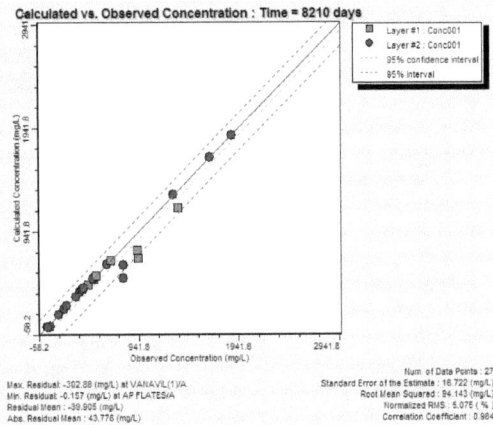

Figure 6.23 Calculated vs Observed Concentration Total Dissolved Solids for 8210 days

6.9.4.2 Total hardness

The prediction of the Total Hardness parameter of the phase of 2015-2016 was made. The corridor stretch from Central to Saidapet showed the safer zone and showed the concentration ranging from 100 mg/l to 400 mg/l and the same value range existed from Central to Anna Nagar.

The end of the stretch showed very higher values in the corridor Central to Tirumangalam and the concentration was in the range 700 mg/l to 1000 mg/l. The Contour lines are not closer enough in both the corridors. The values were predicted when the calculated and the observed concentrations showed the correlation coefficient of 0.995 at 95 % of confidence interval. The calculated and the observed concentrations of the Total dissolved solids parameter for 8210 days are shown in Figure 6.24.

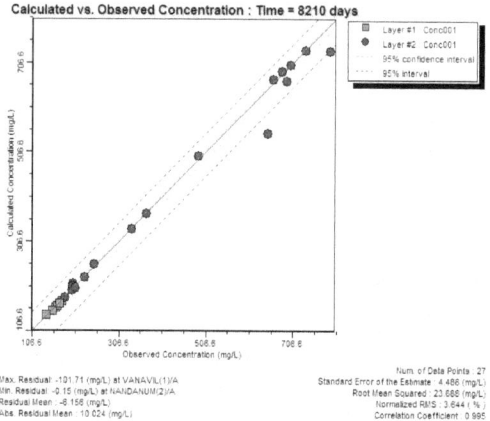

Figure 6.24 Calculated vs Observed Concentration for Total Hardness for 8210 days

6.9.5 Model Prediction for the Phase 2017-2020

6.9.5.1 Chloride

The Contaminant transport of Chloride parameter for the prediction phase 2017-2020 was analyzed. In the corridor from Central to Tirumangalam the chloride contaminant transport was observed in each step of the contaminant transport modeling. During the calibration, validation phases which were segmented with the progress of tunneling work being observed. The concentration seemed to have decreased to the right side of the corridor 2 and the North side of the corridor seemed to be unaffected. In the corridor I the concentration seemed to be progressively increasing from the East to West direction that is almost covering the mid of the corridor I that seemed to be slightly polluted. The concentration seemed to have tremendously increased from Anna Nagar to Tirumangalam with the highest concentration of chloride parameter.

During the calibration period also the chloride concentration seemed to be moving in the same direction from the West to East direction in the corridor 2. In the same way the corridor I also showed the same scenario of moving it from the East to West concentration, but the concentration values were getting high as the progress of tunneling was carried out in the study area. The changes in the velocity direction did not have an impact in the contaminant transport. The Contour maps of Chloride during 2017-2020 and during calibration are shown in Figures 6.25 (a) and (b).The calculated and the observed concentrations are shown in Figure 6.26.

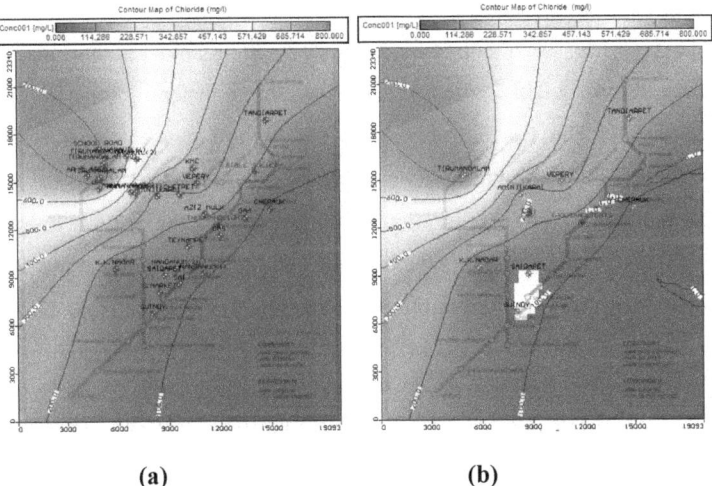

(a)   (b)

Figure 6.25 Contour map of Chloride (a) 2017-2020 (b) During Calibration

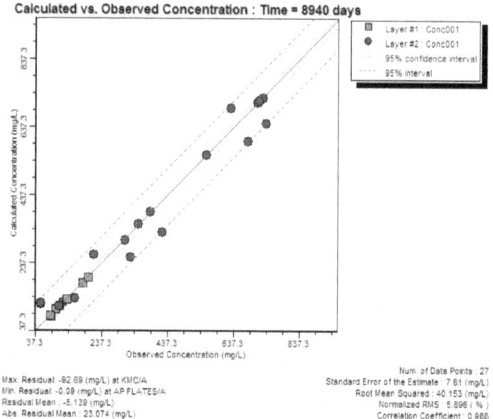

Figure 6.26 Calculated vs Observed Concentration for Chloride for 8940 days

6.9.5.2 Fluoride

The prediction period of 2017-2020 showed that the fluoride concentration highly affected the corridor I when compared with the corridor II. The Corridor I showed the highest concentration range from 0.584 -0.729. The concentration progress was increasing from the South to North direction for the corridor I. In the corridor II the portion from Central to AGDMS seemed to be affected with the highest concentrations that were noted again from the South to North direction. The portion from the AGDMS to Saidapet showed the decreased concentrations and unaffected.

The calibration period and the prediction showed a huge variation in corridor II. During the calibration period the concentration values were very less in each part of the corridor mentioned above in the prediction phase. This tremendous change could be due to the artificial weathering of rocks in the study area. Artificial weathering in the progress of tunneling increased the concentrations of fluoride in the prediction phases. The Contour maps of fluoride during 2017-2020 and during calibration are shown in Figure 6.27 (a) and (b). The calculated and the observed concentrations are shown in Figure 6.28.

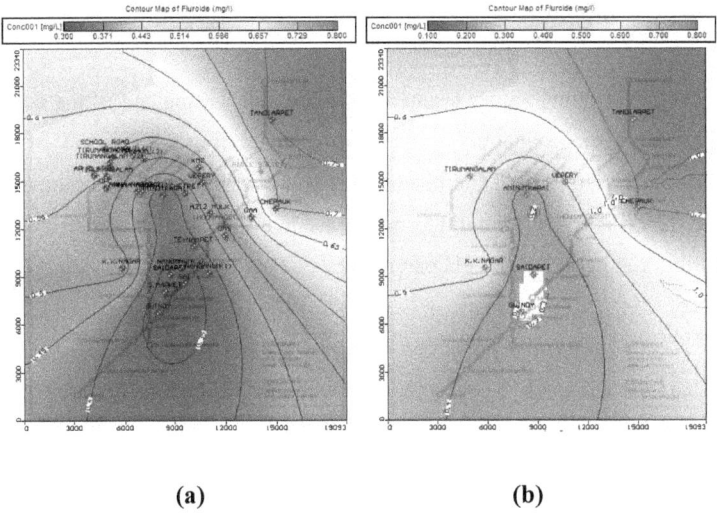

(a)   (b)

**Figure 6.27   Contour   map   of   Fluoride   (a)   2017-2020   (b)   During
Calibration**

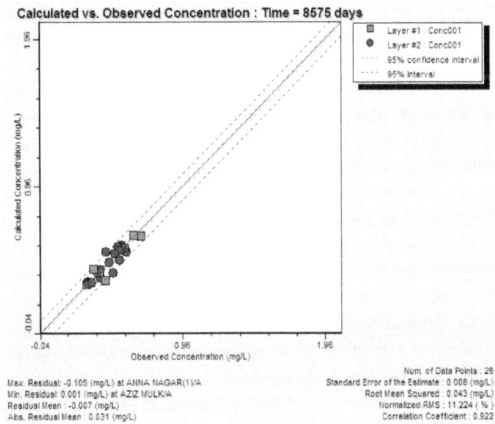

**Figure 6.28   Calculated   vs   Observed   Concentration   for   Fluoride   for
8575 days**

6.9.5.3   pH

The pH values seemed to be low only at the end of the corridor II, that is at the Saidapet region during the calibration period of the pH parameter. The rest of the places, that is from Central to Teynampet area in the Corridor II and the full Corridor I seemed to be highly affected during the calibration period of the pH parameter.

During the prediction phase of the pH parameter the concentration values varied much when compared with the calibration period. The values seemed to be high, almost covering the East side of the corridor II in the range of 7.943 to 8.20 which exceeded the neutral value of the safe water. It clearly showed that the corridor II had got highly affected on both sides with the increased value of above 7.The Corridor I shows that the contaminant moved with the increased concentration from the West to East side of the corridor with values again above 7 but upto the concentration limit not exceeding 7.8. pH parameter seemed to be highly polluted with the construction of tunneling. The Contour maps of pH during 2017-2020 and during calibration are shown in Figures 6.29 (a) and (b).The calculated and the observed concentrations are shown in Figure 6.30.

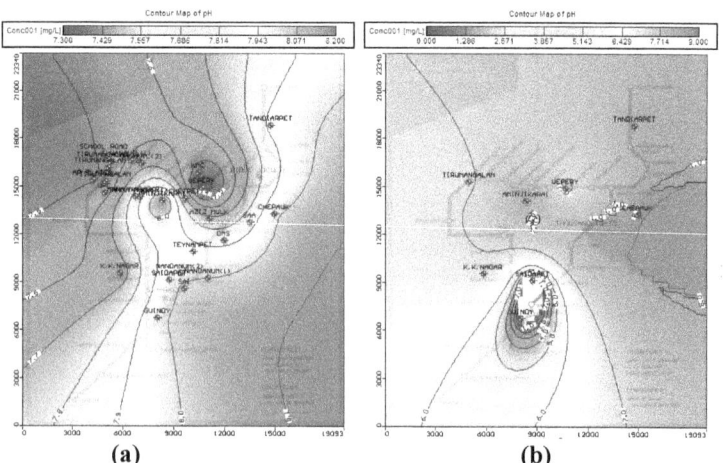

(a)   (b)

Figure 6.29 Contour map of pH (a) 2017-2020 (b) During Calibration.

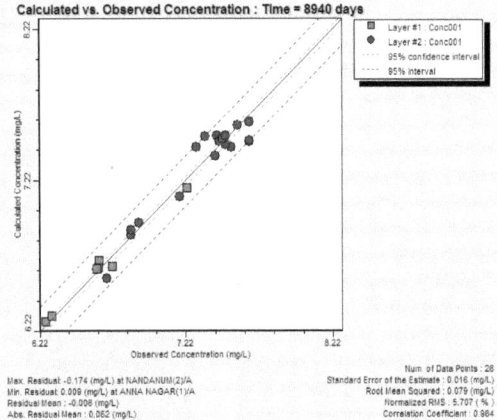

Figure 6.30 Calculated vs Observed Concentration for pH for 8940 days

6.9.5.4   Total dissolved solids

Total Dissolved solids consist of inorganic salts comprising of calcium, magnesium, potassium, sodium, bicarbonates, chlorides, and sulfates and some organic matter that is dissolved in water. The parameter Total Dissolved Solids presented the same scenario in the calibration, validation, and prediction phases. The beginning part of the corridor II that is from the Central to Aminjikarai showed the highest concentrations in the range from 1428 ppm to 2500 ppm. The rest of the places in both the corridors remained unaffected and the values fell in the range from 0 ppm to 1100 ppm.

During the calibration phase the same scenario existed with slightly lower concentrations. It clearly showed that the construction of tunneling had made an impact and caused changes in the concentrations. The TDS level of the fresh water was 500 ppm and for the brackish water 500 to 30000 ppm. Hence the values of the concentrations are likely to be acceptable in both the corridors. The Contour maps of Total Dissolved solids during 2017-2020 and during calibration are shown in Figures 6.31 (a) and (b).The calculated and the observed concentrations are shown in Figure 6.32.

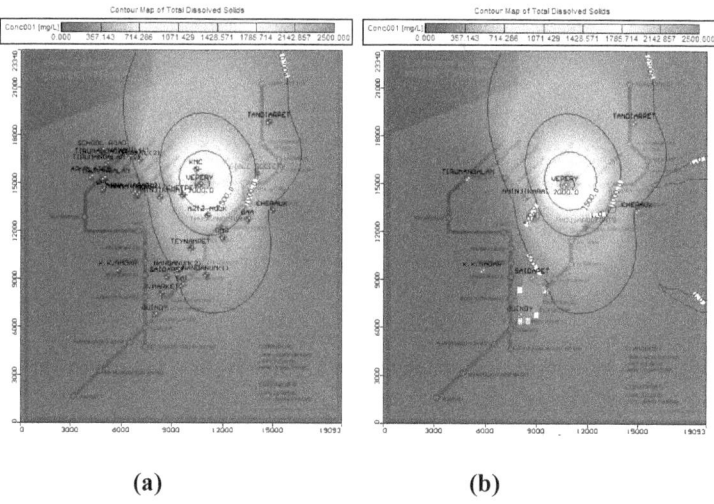

(a)   (b)

**Figure 6.31   Contour   map   of   Total   Dissolved   Solids   (a)   2017-2020
(b) During Calibration.**

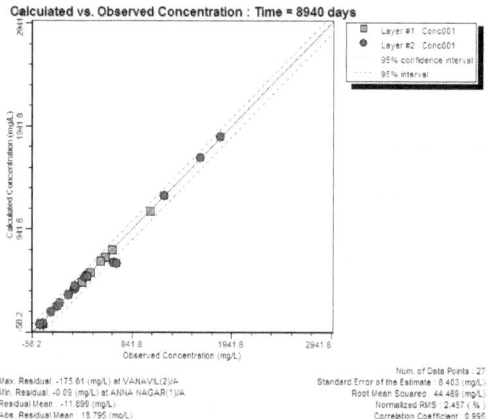

**Figure 6.32   Calculated vs Observed Concentration for Total Dissolved
Solids for 8940 days**

6.9.5.5 Total hardness

The presence of Calcium and Magnesium in water makes the water hard. Total Hardness(TH) does not cause serious health hazards, but it needs softeners to be used in water before use for any application. As the calcium parameter showed the concentration values progressively increasing from the East to West direction, the Total Hardness showed the same scenario of increasing concentrations from East to West direction in the study area. Tirumangalam showed the highest concentration upto 180.00 mg/l and hence it showed the characteristics of very hard water.

The Corridor I on both North and South sides of the corridor showed the range of slightly hard water ranging from 20 mg/l to 88.51 mg/l. The portion from KMC to Anna Nagar showed the moderately hard range of hard water. During the calibration period the same scenario existed for this parameter. The Contour maps of Total Hardness during 2017-2020 and during calibration are shown in Figure 6.33 (a) and (b).The calculated and the observed concentration are shown in Figure 6.34.

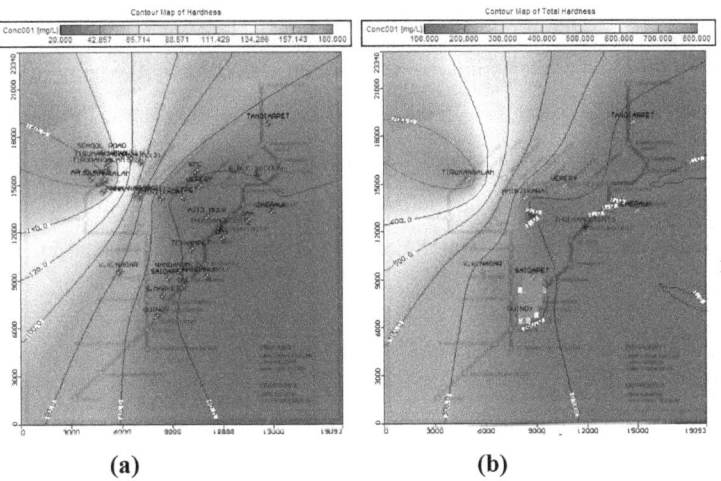

(a)   (b)

Figure 6.33 Contour map of Total Hardness (a) 2017-2020 (b) During Calibration

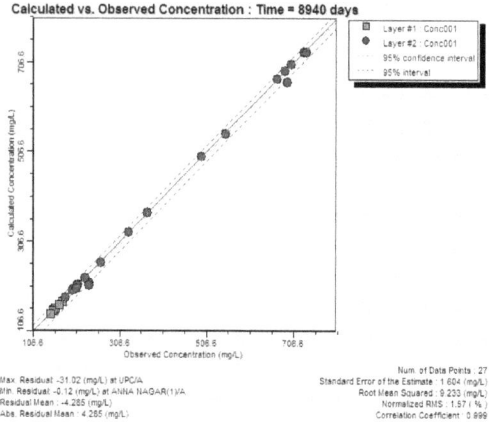

Figure 6.34   Calculated vs Observed Concentration for Total Hardness for 8940 days

6.9.6   Mass Balance of the Water Quality Parameters

Mass balance of the water quality parameters are shown in Figure 6.35. Due to the changes in the physical characteristic of the aquifer, even after the construction the mass which was entering the aquifer was high when compared with before the construction. The sinks out value also seemed to be progressively increased from before to after the construction phase of construction. The reduction in the recharge and sub surface runoff also could be the reason for increase in the concentration of mass in the aquifer.

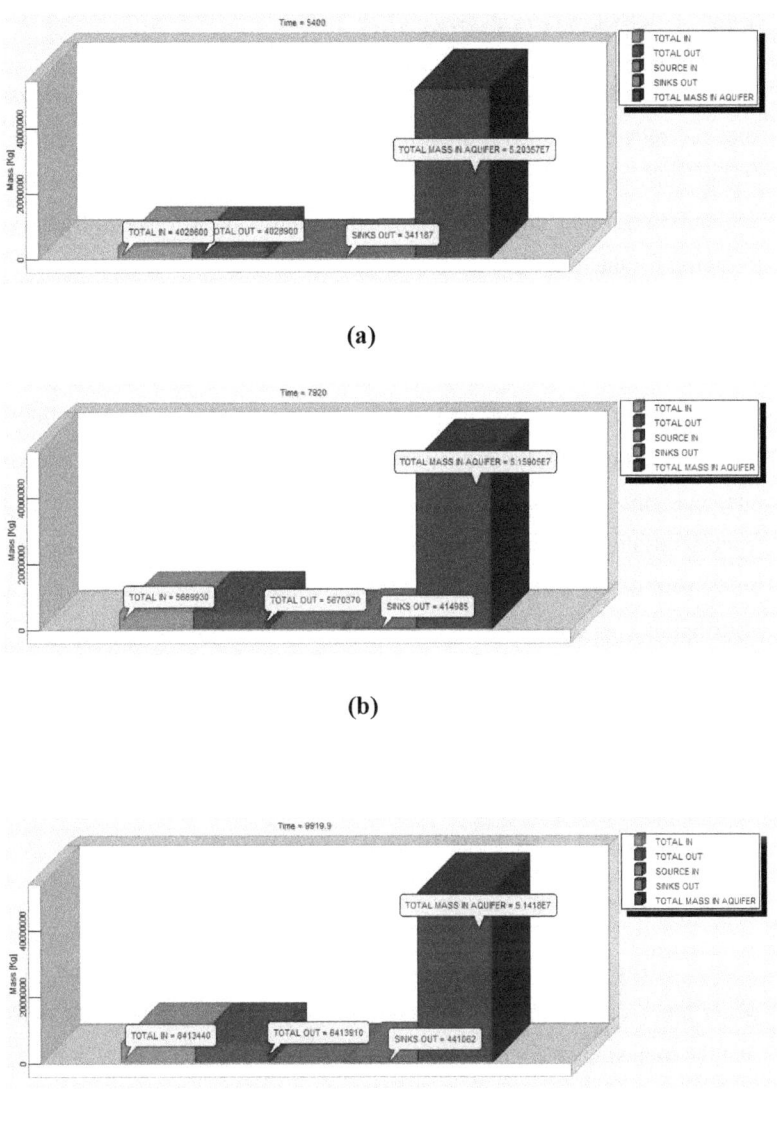

(a)

(b)

(c)

**Figure 6.35 Mass Balance of water quality parameters for the time step
(a) 5400 days (b) 7920 days (c) 9920 days**

6.10   SENSITIVITY ANALYSIS

Within a reasonable range the model input parameters are varied and the relative change is observed in the model response. The process of doing this procedure is called sensitivity analysis. The changes in the hydraulic head and flow rate are noted. A sensitive analysis will give good response for the future predictions.

The selected parameters concentration is a function of conductivity, recharge, dispersivity, and pollutant load. The impact of tunnel construction on the various phases was studied. Conductivity was changed by ± 20 % of the value assigned in the model; at each node the change in the conductivity affected the groundwater velocity causing redistribution of solute concentration. In general, the higher the conductivity, the faster is the movement of the solute and vice versa. Similarly it was observed that head distribution was very sensitive even for 10 % increase in the recharge.

The longitudinal dispersivity and the transverse dispersivity variation do not cause any significant changes in the head and concentration were noticed due to increase or decrease in the dispersivity.

6.11   OUTPUT OF INTERVIEW SCHEDULE

6.11.1   Water level Outputs

The questionnaire for Survey prepared, mainly focused on the water level reduction over the study area. The outputs got from the analysis and the modelings were taken and the questions were prepared with the focus on finding the ground truth of the study. Dwelling units were selected on the observation wells where they were located. On either side of the corridor questionnaire was distributed along with neighborhood areas on either side of the corridor. In the Corridor I, that is from Central to Thirumangalam the responses of the habitats on the North and South side of the corridor are shown in Figures 6.36(a) and (b).

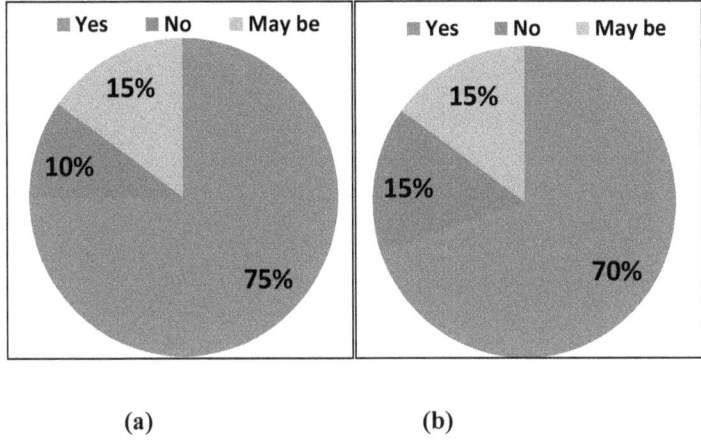

(a)   (b)

Figure 6.36 (a) Response of the North Side of the Corridor I
(b) Response of the South Side of the Corridor I

As per result of the analysis done, the question that was put to the dwellers was whether the water level got increased in the North side of the corridor. Most of the responses about 75%, matched the results obtained. To the South side of the corridor, nearly 70 % of the people agreed with that got in the analysis. In the same way, the other underground corridor, that is from Washermenpet to Saidapet also was subjected to the same methodology and the results are given below in the Figures 6.37 (a) and (b).

The acceptance level of 80% was attained for the question, whether the water level had got increased in the East side of the corridor. Similarly, 76% of acceptance had been obtained for the decrease in the water level on the west side of the corridor.

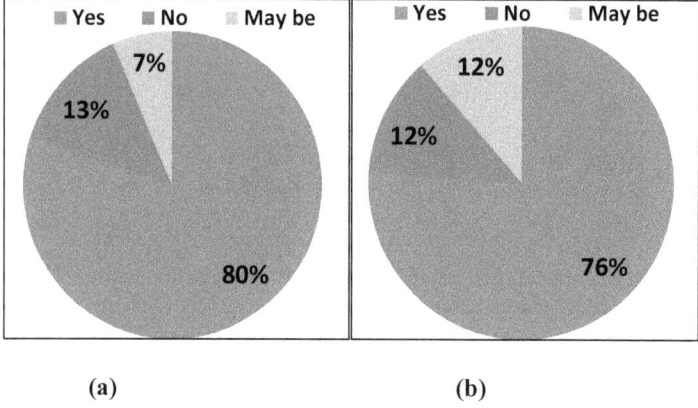

(a)   (b)

**Figure 6.37   (a) Response of the East Side of the Corridor II (b) Response
of the West Side of the Corridor II**

6.11.2   Water quality Outputs

The survey focused not only on the water level reduction but also on
the quality of water around the underground metro rail corridor. Again the
questions were prepared based on the results obtained from the results of
modeling. The response of the people are given under in Figures 6.38 (a) and (b).

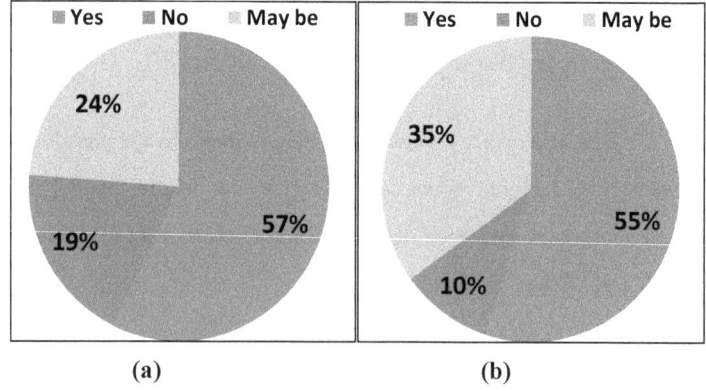

(a)   (b)

**Figure 6.38   (a)   Response   of   the   North   Side   of   the   Corridor   I
(b) Response of the South Side of the Corridor I**

The water quality deterioration could not be seen as much as that of water level reduction and the water quality degradation happened to be a long time process. The response regarding the water quality on both North and South sides of the corridor was about 50 % and it was in the accepted range with the simulated water quality status from the analysis and modeling. Similarly, on the next underground corridor the following results were obtained from the respondents as shown in Figure 6.39 (a) and (b).

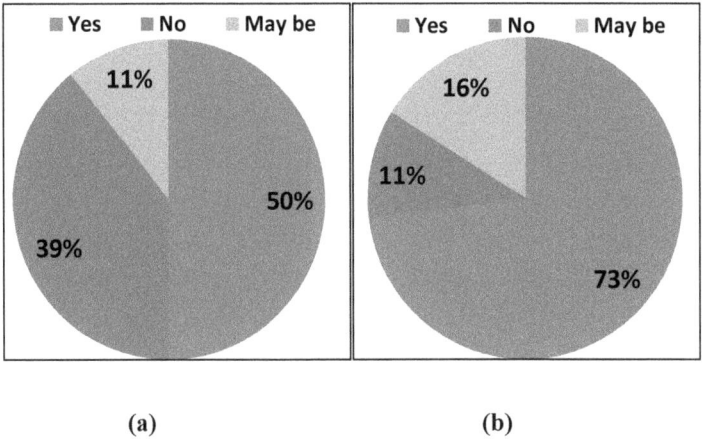

(a)   (b)

Figure 6.39   (a) Response of the East Side of the Corridor II (b)Response of the West Side of the Corridor II

In corridor II, 73% of people accepted that the water quality had slightly decreased. In the East side of the corridor it was only of about 50 % of people who accepted the issue regarding the water quality. The results clearly showed the water quality had got deteriorated where water level had decreased. Always high water level has the higher dissolution than the lower water level side. It has been clearly proved from the above outputs.

6.12   SUMMARY

This chapter summarizes the effect on water quality on either side of the corridor I and corridor II through spatial analysis. The model predicted the changes in the concentrations occurred in the three construction phases of tunneling. The results of the survey is interpreted with the results obtained both for water quantity and water quality.

CHAPTER 7

SUMMARY AND CONCLUSION

7.1   SUMMARY

Chennai is considered as an important city of Tamil Nadu as it provides employment and life line for most of the people living in and out of the city. Hence the city has to meet the needs of the people in terms of water resources and its quality for their healthier lifestyle. To fulfill their essential needs the city is in the position of developing its infrastructure both above and below the ground surface. This kind of urbanization makes the city a developed one and has its own advantages and disadvantages.

In the present research water level reduction and the water quality deterioration due to the massive construction of underground Metro rail corridor was studied. For the research the water level data were collected and even observed in the observation wells. Data regarding to water qualities were collected from the PWD sections and the sampling was done in the laboratory. All these data were separated into two time periods before and after the construction of Metro rail corridor to study the changes occurred after the massive construction.

A holistic and interdisciplinary approach was attempted to attain a sustainable solution. GIS analysis, a graphical representation was used to analyze the data before and after the construction both for the water level and water quality being analyzed. For water quality, water quality index methodology was adopted to study the status of water quality in the study area. Using the available data in the study area a flow modeling and a contaminant transport modeling were done to predict the changes both in the flow and in the contaminant. Different phases of completion of the underground metro rail corridor were considered as the different scenarios for the prediction in the changes of flow and contaminant transport.

In addition to this socio analysis was done in the form of interviews and correlated to the results obtained from the modeling. The outcome of the social and technical analysis was scrutinized and interpreted to suggest the management techniques for the study area.

The water levels seemed to be high before the construction phase in both post monsoon and pre monsoon periods. During the construction phase, a decrease of water level was observed during both the seasons. It could be due to the disturbance created below the ground surface by tunneling.

After the tunneling, during the post monsoon season nearly 40% decline of water level had occurred from the tunneling construction phase. Similarly, in the pre monsoon period decline of water level seemed to be 59% in the maximum values range of 'during' and 'after' the tunnel construction. The excess withdrawal of water during the tunnel construction and also due to the changes in the properties of the aquifer the water level had gone down.

In the spatial analysis, the premonsoon water levels showed that before the construction, the Corridor I stretch from Central to KMC, seemed to be affected with low water levels in the range of 4.0 to 6.0m. In Corridor II, Central to Gemini, the stretch showed the lowest water level range of 6.0 to 8.0m.

During the construction of the corridor stretch I, The stretch that showed the 12.0 to 14.0m range got decreased to 8.0 to 10.0m.Similarly in the corridor II the water level range 12.0 to 14.0m range got vanished again and it was occupied in 8.0 to 12.0m range. After the construction due to the excess withdrawal of water during the construction and also due to the changes in the aquifer properties, almost 50% of the stretch form Central to Thirumangalam the water level got reduced to -1.2 to 2.0m.

Comparing the postmonsoon values, the water level showed the same pattern as that of the premonsoon values but due to recharge in the wells the water levels had been little higher than the premonsoon values.

During the construction because of tunneling, the corridor I showed a small variation in the initial stretch the water level from 8.0 to 10.0m had got reduced to 6.0m to 8.0m.In the corridor II the stretch that showed the water level 12.0 to 14.0m was reduced to 10.0 to 12.0m. After the construction the post monsoon values in the corridor I, Central to Shenoi nagar the water level had reduced to 4.0m to 6.0m. In corridor II, Central to Gemini the water level 4.0 to 6.0m was available and from the Gemini to the end of the underground stretch, the water level of 6.0 to 8.0m existed. The spatial analysis revealed that water level had got reduced after the construction.

In the corridor I during the post monsoon period and the pre monsoon periods South side of the corridor got affected with low water levels range. In the corridor II, that is from Washermenpet to Saidapet, West side of the corridor showed a decreased water levels when compared with the East side of the corridor in both seasons.

The recharge patterns of the wells had got changed after the construction, Tondiarpet well recharge capacity was reduced from 44% to 29%. Similarly, in the wells like Thirumangalam and K.K.Nagar the recharge percentage had got reduced. In Thirumangalam the value dropped from 73% to 60% and in K.K.Nagar it was about 57% to 50%. Wells like Vepery, Saidapet and Aminjikarai showed slight increase in the recharge percentages.

The change in the ground water storage was found using water table fluctuation method for both before and after the construction phases. Specific yield, water level fluctuation and the area of influence of the wells were used to find the storage. The storage before the construction was found to be 0.028 Km^3 . The storage after the construction was found to be -0.128 Km^3. This showed that the storage capacity of the aquifer had gone down after the construction of these corridors.

Comparing the water quality parameter pH during all the three phases during both the seasons it did not show any significant changes. All the values fell under the range of 7.5 -7.9. Total dissolved solids seemed to be slightly increased from before to after the construction that was from 890 mg/l and 933 mg/l to 945 mg/l and 938 mg/l.

Total hardness value did not seem to have changed much in all the three phases but after the construction the values had gone to the range of 776 mg/l in the post monsoon and 770 mg/l in the pre monsoon period. Similarly, chloride concentrations also showed the same trend. It had not changed much in the early two phases but after the construction it showed an increased value of 874 mg/l and 856 mg/l in the post monsoon and pre monsoon periods. But fluoride showed a reduced value of about 0.47 and 0.42 in the post monsoon and premonsoon seasons. Earlier in the before and during the stages of construction, its values seemed to be 0.5 to 0.54.

Before the construction of metro rail corridor, the premonsoon value showed that the pH value around the two underground corridors seemed to be greater than 7.5. During the construction, the pH values in the stretch KMC from Thirumangalam showed less than 7.5 and the rest of the status remained the same. Similarly the post monsoon values before the construction, West side of the study area took the value less than 7.5 and the remaining area registered more than 7.5. After the construction, almost a value greater than 7.5 existed in the two underground corridors.   It clearly showed that the water quality parameters had been greatly affected by the tunneling process.

In respect of total dissolved solids, before the construction, had values less than 500 mg/l in the West side of the corridor and also to the south side of the corridor. The initial stretch that is from Central to KMC in corridor I and Central to LIC in corridor II , carried the higher range of greater than 1500 mg/l. During the construction the area with value less than 500 mg/l seemed to be not existing even in the after phase construction. The postmonsoon values did not seem to be much varied from the premonsoon values in before, during and after the construction phase. It all occurred due to the modification of layer and index properties of the soil.

As regards total hardness, the pre monsoon values showed that before construction all the areas carried the value from 250- 500 mg/l except the stretch, Central to KMC carried more than 500 mg/l in the corridor I. In the corridor II the stretch below Saidapet carried the range below 250 mg/l. During the construction, the stretch from Shenoi Nagar to the end of Thirumangalam became less than 250 mg/l. In the corridor II, from Central to LIC the East side of the corridor got affected with higher values of more than 500 mg/l. But after the construction it could be due to the addition of chemicals containing the composition of both calcium and magnesium, the whole study area carried the Total Hardness more than 500 mg/l.

The post monsoon values also revealed the same status that after the construction the study area got affected with higher values. Use of cementious materials could have increased the values of Total hardness in the study area.

The Chloride seemed to be present more than the permissible limits almost fully in the study area after the construction of corridors when compared with the before and during the construction phase of the corridors.

The water source that was used during the tunneling process and also some of the chemical compounds that were used during the construction could be the source of the chloride and became the reason for showing increased concentration after the construction.

Fluoride concentration seemed to be within the permissible limits. Some areas showed the higher value of greater than 0.83 mg/l but less than 0.95 mg/l. The whole study area founds to be in the range of 0.4 to 0.6 mg/l.

In the flow modeling, the different phases of construction taken were January 2012 – December 2013, January 2014 –December 2015, January 2016 –March 2017 and March 2017-March 2020.Under Scenario 1, the aquifer disturbance was observed and shown in the contour map. The flow was observed from South-West direction for the corridor Central to Thirumangalam. For Washermenpet to Saidapet the flow direction occurred from West to East direction. The Scenario 2 construction showed the velocity direction changes towards the west direction becoming narrower near Vepery. In Scenario 3 the primary observation wells showed the same direction but flow direction became much narrower and straight in the same directions in both the corridors. In scenario 4, after the completion of construction of the stretch the velocity direction completely changed in the corridor Central to Saidapet; earlier the flow direction had been from West to East direction that got changed to South to North direction. The corridor Central to Thirumangalam followed the velocity direction very straight to the West direction. From the mass balance analysis the storage was found to be low in the prediction of 2017-2020.

The contaminant flow modeling of the selected parameters showed no changes in the prediction Scenario 1, Scenario 2 and Scenario 3.In Scenario 4, appreciable changes were predicted in the concentration in both the corridors. In the corridor I, from Central to Thirumangalam pH and TDS showed higher range of values in the North side of the corridor. In corridor II,

Central to Saidapet, the East side of the corridor showed the changes in the concentration, especially chloride and Total Hardness.

The questionnaire survey of the community dwelling on either of the corridor, was taken. The output of the questionnaire survey was overlaid with the results obtained from the water level and water quality status of the study area. About 70% of the respondents view matched the results obtained.

7.2   CONCLUSIONS

Metro rail corridor with massive underground tunneling construction has its own significant effect on both water quantity and water quality. The major conclusions derived from the study are:

- The water level reduction on one side of the corridor happens along with an increase in the water level on the other side of the corridor. The tunneling does not impact any major change in the water quality but progressively it will create a major threat to the water quality in the surrounding areas.

- In corridor I (Central to Thirumangalam), south side of the corridor got affected with low water level. Water level of 2.3m decrease was found in the south side of the corridor when compared to the north side of the Corridor I.

- In corridor II (Washermenpet to Saidapet), west side of the corridor got affected with low water level. Water level of 1.2m decrease was found in the west side of the corridor when compared to the East side of the Corridor II.

- In Spatial analysis, the Corridor I shows a decrease of 2m water level at the end of Thirumangalam after the Corridor construction. Similarly, in Corridor II, the water level range 8m – 10m in Gemini stretch is increased to 10m- 12m after the Corridor construction.

- Overall decrease of 6% in the recharge capacity of the wells was observed.

- Using Water Table Fluctuation method the ground water storage of 0.028 km^3 was found before the construction and the storage of about -0.128 km^3 was found after the construction.

- In water quantity modeling, mass balance storage result shows averagely of about 60% decrease in the storage after the Metro rail corridor construction.

- During the prediction, corridor I showed a decrease of 2.5 m of water level and an increase of 2.1m in corridor II.

- In Water Quality Index analysis, the water quality index value 52.63 was found before the construction and this value was increased to 54.32 after the construction. The index value seemed to be increased after the construction and becomes evident that the water quality is getting deterioted.

- In water quality analysis, the parameter pH shows a increase of 1 in the South side and a increase of 0.4 in the East side when compared to the North and West side of the corridor I and II respectively. The parameter Chloride shows a increase of 150 mg/l in the North side and 200 mg/l increase in the West side of the Corridor I and II after the construction.

- The parameter Total Dissolved Solids shows the value of 210 mg/l increase in the South side and 400 mg/l increase in the East side compared to the North and West side of the Corridor I and II respectively. The parameter Fluoride shows a increase of 0.2 mg/l in the North side and 0.03 mg/l in the West side when compared to the other side of the corridors. The parameter Total Hardness shows a increase of 80 mg/l in the South side and 100 mg/l in the West side was found in the Corridors.

- In spatial analysis, the parameter pH shows the value of more than 7.5 in both the corridors after the construction. After the corridor construction, the parameter Total dissolved Solids shows the range of 500 -1500 mg/l and the parameter Total Hardness shows more than 500 mg/l. Chloride parameter shows more than 600 mg/l and Fluoride parameter shows a increased range of 0.4 – 0.6 mg/l after the Metro rail construction.

- In water quality modeling, the mass balance results showed the increased concentrations in the aquifer. During prediction, pH value increased upto 9 and Total Dissolved Solids increased upto 2000 mg/l. The parameter Fluroide increased to 0.85 and Total Hardness, Chloride increased to 800 mg/l during the prediction.

- The observed water quality change is in line with the groundwater flow direction and water quality variation matches the people's perception.

7.3 RECOMMENDATIONS

The following recommendations to maintain the water quantity and quality around the Metro rail corridor have been suggested:

- The ground water usage on the South side of the corridor I, Central to Thirumangalam and on the West side of the corridor II, Central to Saidapet should be treated and consumed for domestic purposes.

- The erection of more recharge structures and the construction of wells is needed on the reduced water level side of the corridors.

- Optimum pumping rates and the better conservation practices of water management should be carried out in the alloted side of the corridors.

- Spatial and temporal variation of water quality and quantity has to be properly monitored in and around the Metro rail corridor.

- An increased number of observation wells located in the hot spots will be helpful in monitoring the ground water level and quality.

- Sub-surface infrastructure information can be accessed through an Information analysis system that will serve as a readily accessible data management system.

7.4   MERITS AND LIMITATIONS OF THE STUDY

7.4.1   Merits of the Study

- People's involvement throughout the study has given unbiased information for technical and social analysis.

- The modeling of the ground water level, contaminant transport and the prediction has been done with more primary data which clearly depict the ground truth of the study area.

- The study created awareness among the community in and around the metro rail corridor regarding the water level and water quality.

7.4.2   Limitations of the Study

- Adequate data were not available regarding the aquifer characteristics and hence suggestions on appropriate recharge structures to overcome the water level reduction   could not be made and it is a significant limitation.

- The study could not establish a linkage between health issues and the increase or decrease of the parameters taken for the analysis due to time constraint.

7.5   CONTRIBUTION OF THE PRESENT STUDY

- So many underground developments are made throughout the world to improve comfort of people. Its impact on water resources need to be analysed so that improvement on one aspect should not lead to deterioration of another, sustainability should be ensured.

- This study attempts to find the impact on water quantity and water quality due to this underground development. Effect on storage, changes in the water level and the changes in the concentration were analysed.

- Groundwater modeling was done with the primary water level data with the cooperation of the people to find the exact scenario of the water movement below the ground surface due to the tunneling.

- The technical findings from the modeling were verified with the ground truth along with the ground water flow and the people's perception regarding the water level reduction and water quality. This interdisciplinary approach proves to be fruitful for future research.

- This thesis emphasize its useful contribution towards to the occurrence in terms of water level, its movement in the form of velocity directions, transport of solutes, physical and chemical properties of water thus covering the overall hydrological components of the environment.

7.6   SCOPE FOR FUTURE STUDY

The following are a few areas for further research:

- The potential of sea water intrusion in the coastal side to the fresh water interface can be predicted due to the sub surface development.

- Hydrology information accountability like water budget research for different management scenario can be done for the restoration process in the aquifer.

- The available techniques can be optimized and a better remediation technology may be developed.

- Groundwater contamination linkages can be identified to avoid health problems and can be formulated.